D0758244

FastLane

JOHNS HOPKINS STUDIES IN THE HISTORY OF TECHNOLOGY

Merritt Roe Smith, Series Editor

FastLane

MANAGING SCIENCE IN THE INTERNET WORLD

Thomas J. Misa + Jeffrey R. Yost

Johns Hopkins University Press | BALTIMORE

© 2016 Johns Hopkins University Press
All rights reserved. Published 2016
Printed in the United States of America on acid-free paper
9 8 7 6 5 4 3 2 1

Johns Hopkins University Press
2715 North Charles Street
Baltimore, Maryland 21218-4363
www.press.jhu.edu

Library of Congress Cataloging-in-Publication Data
Misa, Thomas J.
FastLane : managing science in the Internet world / Thomas J. Misa, Jeffrey R. Yost.
pages cm. — (Johns Hopkins studies in the history of technology)
Includes bibliographical references and index.
ISBN 978-1-4214-1868-1 (hardcover : alk. paper) — ISBN 978-1-4214-1869-8 (electronic) — ISBN 1-4214-1868-1 (hardcover : alk. paper) — ISBN 1-4214-1869-X (electronic) 1. Science—Data processing. 2. Computers—History. 3. Internet in public administration. I. Yost, Jeffrey R. II. Title. III. Title: FastLane. IV. Title: Managing science in the Internet world.
Q183.9.M57 2016
502.85—dc23

2015010637

A catalog record for this book is available from the British Library.

Special discounts are available for bulk purchases of this book. For more information, please contact Special Sales at 410-516-6936 or specialsales@press.jhu.edu.

Johns Hopkins University Press uses environmentally friendly book materials, including recycled text paper that is composed of at least 30 percent post-consumer waste, whenever possible.

Contents

Preface

This book has its origins in an offhand remark made by Bruce Seely, who at the time was serving as a National Science Foundation (NSF) program officer in science and technology studies. As a temporary program officer, or "rotator," Bruce looked around him during 2000–2002 and saw firsthand the changes that NSF was experiencing with FastLane. It was a new computer system that ran the central mission of NSF: making grants to the nation's research community. "FastLane is fundamentally changing NSF," he observed, "but NSF doesn't have insight into *how* it's happening—or *why*." His comment had staying power and unusual gravity, for here were two vital forces in contemporary society—science and computing—obviously bound together in an immense and consequential real-time experiment.

The authors began working together during the spring of 2006 at the Charles Babbage Institute (CBI) at the University of Minnesota. Yost had arrived several years earlier as associate director of CBI and was well versed in computer history. Misa was incoming CBI director. In March that year he went to a workshop at Indiana University, where he had the good fortune to meet NSF's Suzanne Iacono. Over lunch, Iacono promptly agreed with Seely's diagnosis about the importance of FastLane and the need for insight.

In developing a proposal, Misa made a visit that fall to NSF in hopes of locating research materials on FastLane. Traditional paper documents about FastLane were not readily available. But he had luck in interviewing eight NSF staff members, all of whom shared specific ideas about shaping such a research project and candid personal assessments of FastLane itself. Our thanks to Dan Atkins, Edward Hackett, Fred Kronz, Daniel Newlon, Robert O'Conner, John Perhonis, Tom Weber, and Fred Wendling. Phone conversations with Bruce

Seely and Connie McLindon reinforced our emerging plans to make a full-scale assessment of NSF and FastLane.

We needed to design a study that would capture the process of design and development of FastLane; that would spotlight the crucial role of FastLane's users, at several levels; that would turn the seeming dearth of paper documents and the evident enthusiasm of NSF staff to share their experiences into a robust research plan; and that would capture the multidimensionality of computer system development occurring in the U.S. federal government. In the back of our minds was the question of the digital revolution: how was it that computing was changing society and politics, identities and institutions? FastLane might be an ideal case to explore these themes. It was a tall order, and a December proposal deadline loomed large.

We crafted a research proposal to NSF, submitted it through FastLane, made the deadline, and then—like countless principal investigators (PIs) beforehand and since—waited for the assessment of merit review and the decision of a program officer. We were fortunate in receiving support from NSF for this study of NSF. A word about our relationship with NSF might be in order. Connie McLindon, based on her deep experience with federal research management (her key role in shaping the immediate forerunner to FastLane is examined in chapter 3), recommended that we conduct this research as a grant rather than under a contract. That way, she correctly indicated, the results are entirely the investigators' and the agency cannot influence their findings in any way. This indeed was our experience with NSF. We had the good luck to find program officers in NSF's Human Centered Computing program who understood the point of this study and arranged a Small Grant for Exploratory Research (IIS 0747445) to get the project off the ground and then, with a revised proposal, a regular grant (IIS 0811988) to fully support the research. Wayne Lutters and William Bainbridge were our program officers. We started work in 2007, spent four years doing intensive research including interviews across the country, filed our final project report in November 2012, and then completed work on this book manuscript.

Our first thanks must go to the more than 400 interviewees who made room in their busy lives to address our in-person questions about FastLane. We are equally indebted to the 400 additional people who took the time to complete our online interview questions. Together, they allowed us to create an unusually large and diverse dataset about the design, development, and use of a computing system. NSF staff were uniformly helpful and unfailingly supportive. We recorded interviews with members of the team responsible for designing, developing, and training for FastLane, including Jean Feldman, Constance McLindon, Carolyn Miller, Andrea Norris, Craig Robinson, Rich Schneider, Beverly Sherman, George Strawn, and Frederic Wendling. Special thanks to NSF historian Marc Rothenberg, who helped us understand NSF's internal dynamics and culture. Documents now in his office's possession are the best written sources on FastLane. At CBI we worked with research assistants Jonathan Clemens, Siddhartha Shanker, Joshua Welsh, and Joline Zepcevski. Katie Charlet, CBI's administrative assistant, took capable charge of the mountain of oral histories: recordings, raw transcripts, edited versions, permission forms, and now the electronic archive. At the University of Minnesota, we had the support of the superb grants-administration staff and especially the attention of Gina McCabe and Julia Sytina.

Our assessment of FastLane is overall a positive one. Yet as the publisher's external reviewer for this book pointed out, our portrait gives voice both to the many supporters who praised FastLane and to the critics who saw its shortcomings—or the alternative paths that might have been taken. Some of these paths might yet be realized. We assembled a set of summary findings and lessons learned in chapter 7. As users of FastLane now for nearly two decades, we strain to recall the age of paper proposals. Perhaps we can leave our readers with this image, from an NSF program officer who responded to our online interview with a recollection of the paper covers, or "jackets," that formerly wrapped hard-copy NSF proposals: "One remembers the pre-FastLane era, memorably, as 'Stacks of jackets. Long hours in the office with stacks of jackets.' "

FastLane

1

Managing Science

Science and computing press forward, seemingly without limit or liability. In a heartbeat of historical time, the enterprise of computing has transformed the world's economies and cultures. It once took computers the size of a dentist's office and costing millions of dollars to do the multitudinous number-crunching that yesterday's desktop computers or today's tablets do in the background while processing incoming streams of audio or video. A torrent of information unimaginable a generation ago can be accessed today, one way or another, by around 40 percent of the world's 7 billion people—among them a billion Internet subscribers in Asia alone—and the number climbs with each smart phone that makes its way to a village in Africa.[1] Companies that have exploited these changes, such as Amazon, Google, and Facebook, have become dominant economic actors in the West, while China is being remade by its own native e-commerce giants such as Baidu, Alibaba, and Tencent. Sometime during 2014, China, with more than 600 million Internet users, likely surpassed the United States as the largest online economy.[2] Countries and cultures that hold themselves aloof are increasingly isolated islands in the digital stream.

Science once explored the natural world of atoms, plants, and solar systems, but we now look to the enterprise of science for breakthroughs in genetics, new materials, nanotechnologies—and computing. Science itself depends ever more on information, especially such data-rich disciplines as astronomy, high-energy physics, brain imaging, evolution, and other branches of the physical, biological, and social sciences, as well as the emerging digital humanities. Many sciences, as artificial intelligence pioneer Herbert A. Simon anticipated in 1969, are now sciences of the artificial. For these, there is no longer a "reality" out there in the natural world; their objects of study are artificial constructs of humans. We depend on these constructs more than we might realize. Information theory emerged from the shadow of wartime cryptography to become a mainstay of computer science and the underpinning of algorithms that detect consequential bits of news from the immense oceans of financial data and instantly trade blocks of stocks. Anyone watching a digital movie or listening to digital audio depends on an information science that has no "real" object. Even exact paper copies can go awry, as Xerox in 2013 found to its dismay when a German computer scientist uncovered an eight-year-old bug in the image-compression software used in 20 models of its copier-scanners that scrambled the images of numbers while making "original" copies.[3] Science is now suffused with virtual and augmented reality models that elaborately simulate and visually project "natural" processes such as those inside distant stars or in close-at-hand watersheds.

The burgeoning computational sciences and their powerful simulation models have remade what it means to be a practicing architect, chemical engineer, or solid-state physicist. Then consider the immersive, but hardly real, experience of movie graphics and computer gaming. From the virtual violence of Grand Theft Auto, which recently smashed all previous records by ringing up $800 million on its *first* day on sale, to the actual computer-driven battlefields of the military, innumerable hours are spent by humans and software bots alike absorbed in a reproduction of reality that has no original.[4] Images—and sounds—from tiny proteins as well as from immense galaxies give scientists new tools for comprehending the natural and manmade worlds. And just when we have nearly grasped these epoch-making developments, along comes the tsunami of "big data."

In this book, we investigate what happens when these two powerful forces—science and computing—are joined. Since we live in a world significantly shaped by computing and science, we ought to know something about their intentional marriage and the offspring we live with today. Our extended case study in this book—the computer system that runs the National Science Foundation (NSF), the preeminent funder of basic science in the United States—is scarcely the first time that science and computing joined together to remake the world. And NSF's FastLane computing system, taking form at the start of the web-based Internet era, is not the last such marriage.

In its treatment of computing, this book focuses on the years surrounding the birth of the World Wide Web. By the mid-1980s the Internet was emerging as a "network of networks" linking universities, government agencies, selected private and public computer networks, and countries around the world, initially in a noncommercial environment. Military researchers and the nation's elite universities had already been connected through ARPANET, which had approximately 70 nodes linked by high-speed telecommunication lines, a hundred or more local host computers, and a few thousand network users by 1980, when, arguably, the first virus hit it. On 27 October 1980 an errant status message spreading across the network overwhelmed the network's servers and temporarily brought down the entire ARPANET (see figure 1, p. 90). Other network users at the time explored early offerings on BITNET, USENET, and UUCP while many computer scientists used the telephone-based CSNET. In Europe there were several X.25-based public networks as well as the 20-node CYCLADES network based in France. Indeed, it was for this profusion of diverse networks that Bob Kahn and Vint Cerf famously devised the Internet protocols, soon to become the ubiquitous TCP/IP that spread rapidly in the 1980s. The for-profit Internet emerged a decade later, after the National Science Foundation that ran the high-speed Internet "backbone" called NSFNET (see figure 2) relaxed its rules in the mid-1990s and a host of profit-minded "Internet service providers" flooded in, setting the stage for the dramatic expansion of the World Wide Web and the zest and zaniness of the "dot com" bubble.

The story we tell in chapter 3 begins at the moment of creation: the web-browsing software that was mass marketed as Netscape and as Microsoft's Internet Explorer. We show that the University of Illinois computer science laboratory creating the Mosaic browser—the direct model for Netscape and Internet Explorer—crucially depended on NSF funding and, what is more, that in requesting specific features, FastLane staff shaped the emergence of this prototypical web browser. The story is not known outside NSF in large measure since the University of Illinois instituted legal proceedings that sought complete proprietary control over Mosaic and consequently shaped much of the publicly accessible knowledge. The university, thanks to the Bayh-Dole Act of 1980, already controlled key patents on Mosaic and launched the legal action to prevent the former student developers, by then in Silicon Valley, from using the trademarked name.[5] Along the way we also describe other vital elements that came together in the first generation of e-commerce, e-government, and NSF's own FastLane—including local-area networks such as Ethernet, remotely accessed databases, graphics-rich browsers, and distributed networks of servers.

We show that FastLane's designers were among those who were pioneering in e-commerce and e-government. The two had more in common than we typically remember. It is no accident that the FastLane programmers used the very same software tools (Perl and C) to couple a database securely on a server to the front end of a web browser as did the programmers at Amazon.com, who were creating the prototype for e-commerce in the *same months* as FastLane was taking form. Amazon states that its first online book sale was in July 1995; NSF, that its first FastLane transaction (the submission of a proposal review) was in March 1995—four months earlier.[6] We interviewed one of the FastLane software developers who wrote the computer code that accomplished the feat of connecting a web browser and database (see chapter 3). In just recognition of its pioneering role in e-government, FastLane received a notable National Information Infrastructure Award in 1996.[7] Anyone going online today—to look up best-selling books, check bank balances, or purchase an airplane ticket—uses these protean creations.

The computing model universally used in e-commerce today owes quite a lot to the story we relate in these pages. If the original World Wide Web put reams of data "on" the Internet, freely available and commonly addressed through static URLs, the e-commerce model behind FastLane and Amazon took data "off" the publicly addressable network, owing to legal, privacy, and security concerns, instead storing data in password-protected databases.

Our fine-grained historical treatment of these momentous changes—based on 400 in-person and 400 additional online interviews with FastLane's designers and developers, as well as legions of the system's users, in addition to an extensive review of documents[8]—permits a more incisive and subtle view of technology development than the typical "before and after" snapshots in the emergence of computing. Sometimes when we see the immense gulf (for instance) between a 1970s mainframe computer and a 2000s networked personal computer, we are lulled into thinking that computer *technology* itself is the most important change. Such a "snapshot" viewpoint leads naturally to some version of technological determinism: the assumption that changes in technology drive the big changes in economics, politics, and culture. We imagine that computing changes the world. But this is far from a complete view.

Our book sets the seismic changes in networking, database, and communications technology that FastLane depended on alongside key organizational, regulatory, and cultural changes: these often created the context that led to the technological changes. Fax machines, computer modems, and bulletin-board computer services, for instance, depended crucially on regulatory changes starting with the Carterfone decision of 1968 (13 FCC 2d 420), which opened up the telephone network. Before then, only AT&T's phones were permitted on AT&T's telephone network. The Carterfone decision allowed not only fax machines but soon enough computer modems that made possible dial-in access to large computers or bulletin-board systems running on smaller computers. Author Howard Rheingold coined the term "virtual community" to describe this use of dial-up modems and bulletin boards in the San Francisco area. Absent the Carterfone decision, it's difficult to imagine the vibrant telephone-based online communities that sprang up even before AT&T was restructured in 1984. (One of us used a dial-in telephone modem to access FastLane through 2005.) Determined advocates of technology who make consequential investments in risky, unproven, or experimental technologies—thereby making them into successful technologies—is another recurring theme. And, sometimes, there is the elusive play of contingency and fortuitous accidents. Overall, this book seeks to understand both the shaping of emergent technologies, such as the ones creating today's ubiquitous e-commerce, and the consequences of their deployment, which were especially notable in the policies and practices of science.

In its treatment of science, this book tells a somewhat longer story, focused on the National Science Foundation and its mission-specific needs for information. NSF was created in 1950 and across subsequent decades grew to become the nation's preeminent funder of basic research in science and engineering. To understand how FastLane automated the flow of information at NSF, we investigate and describe how the agency conducted its evaluation of proposals before the arrival of the digital era. We explore the origins of peer review, the effects of congressional oversight, and the changing practices and priorities in the sciences themselves. There was a simmering debate within NSF whether FastLane ought to mimic the existing paper-based proposals or, rather, adopt a reengineered form. In the main, NSF elected to stick with the "paper paradigm," as can be readily seen today. One challenge for NSF in the years to come will be readjusting its existing proposal paradigm with emerging models in the sciences that are simulation-driven or rely on multidimensional motion-picture graphics.

The link between computing and science is FastLane: it forms the central nervous system of the National Science Foundation. Since it became mandatory across the agency in October 2000, FastLane has been the principal information conduit that connects NSF to the worldwide community of researchers and their institutions. FastLane creates an indispensable pipeline for proposing, reviewing, funding, managing, and reporting on the nearly $7 billion that the agency awards in research funding each year. You cannot secure funding from NSF without using FastLane.[9] As such, it is the single most important computer system for the nation's community of basic science and engineering researchers. It is, to slightly bend a phrase from actor-network theory, a quintessential "obligatory point of passage."[10]

Today, it is difficult for many researchers, administrators, and even NSF staffers to remember what things were like prior to FastLane. As recently as the 1980s, scientists might obtain a research or educational grant from NSF using little more than a mechanical typewriter and a suitable supply of Liquid Paper or Wite-Out to draft their proposals. Someone at their university or research institute reviewed and approved the proposal, then made 15 or more paper copies with a standard copying machine before mailing the bulky package to NSF. Four people we spoke with remember using mimeograph machines, a technology dating back to the 1920s. While researchers in the Washington, DC, metropolitan area might hand deliver their proposals to NSF at the last minute before a deadline, researchers in more remote areas such as Alaska and Hawaii, owing to the differences in time zones and the thousands of miles that separated them from the capital, were forced to mail in their proposals at least two days early. A mountain of paper converged on NSF with each proposal cycle—an annual equivalent of perhaps 50 million pages. Today, NSF funds around one-fifth of the 50,000 proposals it receives each year.[11]

FastLane is inseparable from the Internet, just like many computer systems that have become essential to our routines of work and life in recent years. As noted above, FastLane itself took form in the early 1990s just when the Internet as we experience it today was in the formative stage. Even though some individual NSF staff were keen on using computers to streamline their work within the agency, it was not clear how—or if—a universally accessible interactive computer network might be built to connect the agency with the far-flung national community of researchers.[12] Computer science researchers had been sending files over the military-supported ARPANET for more than a decade, but it was by no means clear that a chemist in Montana or a neuroscientist in Puerto Rico would have ready access to a reliable and robust computer network that would determine the fate of their research careers.

While the Internet was an obviously promising place to start, its chief use in the early 1990s was for text-only FTP or telnet logins to central servers and text-only email. Even though Tim Berners-Lee had invented the World Wide

Web in 1991 at a Swiss high-energy physics laboratory (discussed in chapter 3), the Internet that we know today simply did not exist. There was no obvious way using computer networks to send such complex documents as 100-page research proposals containing text, tables, and images, not to mention legally consequential budget figures. Email attachments with binary documents, such as images, graphs, equations, or word-processing documents, as commonly used today, scarcely existed outside of specialized research communities. There was a bewildering variety of experiments with multimedia document standards, such as the EDI standards that had their origins in military logistics and were gaining dominance in the automobile industry, and of course there were business-based networks of telex and fax machines that transmitted paper documents across phone lines. But at the time no one considered the Internet as a means to reliably and securely deliver thousands of documents, sent in from hundreds of universities spread across six time zones and two oceans stretching from Hawaii to Puerto Rico.

As we show in this book, FastLane grew up with the part of the Internet that became the World Wide Web. We tell the story of how NSF staff learned about software created at the University of Illinois by NSF-funded computer-science researchers. In time, their Mosaic web browser and its direct descendants effectively became FastLane's required front-end user interface, as well as the required front-end for nearly all of e-commerce (see figures 4 and 6). Another piece of computer software that NSF drafted to the cause for creating complex documents was Adobe Systems' Portable Document Format (PDF), which at the time was an experimental format familiar only to computer experts.[13] NSF's early adoption and nationwide use helped make PDF into the ubiquitous format that it is today. Yet another now-pervasive computing technology that grew up alongside FastLane was web-linked computer databases where NSF could receive, store, and retrieve the documents comprising a researcher's proposal. (The software that permitted web browsers to access databases was Common Gateway Interface [CGI], as explained in chapter 3.) In short order, users of FastLane learned to use web browsers such as Mosaic, Netscape, and Microsoft's Internet Explorer to interact with NSF databases and much else besides. These elements combined into a new informational paradigm that reshaped e-commerce and e-government.

In effect, you might say that for FastLane NSF was building an immense electronic postal system. There were a number of challenging problems. Literally billions of dollars in research funding depended on the secure transmission of proposal documents: fast and certain registered delivery, you might say. Unlike a regular company, which might simply mandate a certain computer system across all its offices or even impose a particular accounting system, NSF had, in addition to its 1,200 or so professional staff members in Washington, DC, an active community of perhaps 100,000 researchers scattered across the globe. Some NSF-funded researchers had access to the largest and fastest supercomputers in the world (see figure 3); others did not yet have personal computers in their offices. For that matter, it hadn't been so many years that NSF program officers all had computers on their desks.

From the start, FastLane took form with unusual attention to equity and open access. As one NSF visiting-committee member recalled it, "Some colleges, especially some of the historically black schools or Native American tribal colleges . . . didn't really have the Internet. So to require FastLane was considered a hardship."[14] Making FastLane into a *required* system entailed extending the Internet to become a universally accessible resource, and this too is part of the story we tell in these pages. In addition to tribal colleges and historically black colleges and universities (HBCUs), NSF also focuses attention on underresourced states and territories as part of the multiple-agency EPSCoR (Experimental Program to Stimulate Competitive Research) initiative.

FastLane was a huge real-time experiment with what computer researchers were calling "value laden design," where certain organizational values were explicitly designed into computer systems. NSF clearly intended FastLane to express and embody the agency's values of security, interoperability, reliability, and sanctity of merit review. Each of these values was discussed and articulated as the FastLane system took form in the 1990s (as chapters 2 and 3 recount in some detail and chapter 7 revisits and evaluates). A strong preference for user-centered design also was pervasive in these formative years. NSF saw its community of funded researchers not merely as passive users, who might or might not appreciate the new way of doing things; instead, it understood that these users were also the key constituency that the agency needed to keep in mind—ideally to please. How NSF created, deployed, and managed the interactions with users is a distinctive part of FastLane. These chapters in the FastLane story may inspire software engineering and human-centered computing researchers with a notable success story: we tell it from the inside.

Methods and literature

There are several vantage points to understand the emergence of a far-reaching computer system such as FastLane. Historians have long been considering the interaction of technologies with the societies and cultures in which they are embedded. Scholars from several disciplinary backgrounds have increasingly devoted close attention to such technical infrastructures as electricity, energy, transportation, communication, and information. Computer science researchers in software engineering and human-centered computing also can find useful models and insights in this story.

This book adds to our understanding of NSF as a federal research agency shaping science and engineering. It is distinctive in examining research *practices* (many studies of federal science and engineering are limited to research policies) all the way from NSF's origin in 1950 through to the first decade of the new millennia, whereas existing historical work on NSF says little beyond 1990.[15] And because computerizing NSF's core mission of grant making engaged the entirety of the agency, our history of FastLane is an especially revealing window into the agency as a whole. Based on extensive interviewing with NSF staff, as well as scientists and research administrators at 29 universities, the book offers a unique view of the management of science in the Internet age.

FastLane may be something of a model for managers in the federal government and other complex organizations to learn how NSF designed and deployed this new computer system. We hope the possibilities of improvement might resonate with citizens who are displeased that only 1 in 20 large government computing projects are deemed wholly "successful" while fully 40 percent end in outright "failure," such as the $170 million FBI Virtual Case File system, an unworkable networked scheme for tracking criminal cases, abandoned in 2004, or the $1 billion Air Force supply management system known as Expeditionary Combat Support System that was canceled in 2012. Computer problems during the 2014 rollout of the Affordable Care Act in Minnesota and Oregon alone likely wasted $400 million.[16]

As with most all computing technologies, FastLane is best understood as a complex, interacting system of technologies. Some time ago, when historians examined technologies they tended to focus on individual artifacts such as automobiles or light bulbs or the famous inventors such as Henry Ford and Thomas Edison. Scholars examining technology now recognize that a more fundamental object of study is the *system* of technologies that these individual artifacts are a part of, as well as the societies and cultures in which they are embedded. So while the earliest historical studies of computers (for instance) debated who should be considered the pioneering inventor of computing, scholars now are more interested in the business and government *institutions* that supported the development of computing, as well as the economic, social, and political *processes* that shaped and guided these developments.[17] It is not possible to meaningfully answer the question of "who invented" FastLane because there were many people—at least a dozen lent their technical insights, programming skills, and organizational savvy—who contributed to its origin, early design, and subsequent development.

As a system of information technologies, FastLane is an exemplar of how individual technologies are combined to perform useful functions. While individual artifacts, such as browsers, computer networks, databases, and laser printers are obviously important, the true challenge of FastLane was assembling these components into an efficiently functioning system. Curiously, even laser printers were crucial because when FastLane was fully launched in October 2000 all proposals arriving electronically at NSF were first printed out on paper before circulating, as before, within the agency. The agency's internal e-Jacket system took additional years to design, develop, and implement. There were many important points of NSF policy and innumerable daily procedures that needed to be addressed by NSF staff and research administrators across the country.

The emerging group of interdisciplinary scholars who are studying *infrastructures*, such as highway systems or electricity networks, extends the notion of what constitutes technological systems. A Boeing 747, in this view, "is a complex software and hardware infrastructure containing 6.5 million lines of code distributed across dozens of different computing resources."[18] Studies of infrastructures frequently emphasize two additional points. First, while infrastructural technologies obviously go through a stage or period of invention and development, just as all technologies do, the studies of infrastructures tend to emphasize their subsequent embedding in society and consequently deal with a much longer historical time horizon. While historians of technology frequently have focused on the early emergent phase in a technology, infrastructure scholars might extend the temporal window to examine an infrastructural technology across many years or even many decades. This temporal difference means that while historians often stress the flexible or malleable or emergent aspects of a given technology (such as differing voltages for residential electricity distribution or varied line standards for television broadcasting), infrastructure scholars might highlight instead the durable or structuring aspects of an infrastructure (such as the changes brought about by electricity networks in times of working and places of living), a second notable point of emphasis.

Of course, computing at varied points in time can be emergent and flexible or long-lasting and society-shaping. To suggest just one society-shaping example, ever since the 1960s emergence of the ASCII standard character set—that is, the *American* Standard Code for Information Interchange—computers around the world were effectively "hardwired" to be hospitable for English-speaking users and, as a result, inhospitable to users of any other language needing special accents or a non-Latin set of characters. Remarkably, the Internet's reliance on ASCII for its address space carried this 1960s bias, which had its roots in the short length of computer "words," all the way until 2010, when—for the first time—Internet addresses could use non-Latin character sets, such as Arabic or Chinese. In the present study, we combine the attention of historians to the origins and emerging qualities of computer-based infrastructures like FastLane with a complementary awareness of the longer-term consequences and structuring aspects of such a system. To give one instance, our interviews with the pioneering programmers for FastLane make abundantly clear how and why an odd "fork" in software was built into FastLane and why, today, owing to this fork, it takes three teams of programming experts to make any and all significant updates. FastLane, once new and novel and emerging, now consists of highly complex "legacy" code. Whereas the shorter-term perspective shows vividly how social choices are built into emerging technologies, the longer-term perspective can help spotlight social and cultural changes brought about through technologies.

Computer science researchers, especially in the fields of software engineering and human-centered computing, are vitally interested in careful assessments of computer systems. Software engineering researchers may find a model in the sophisticated design thinking that NSF managers were able to deploy, especially the managing of substantial software uncertainties, the early attention to buy-in by different types of end users, and, most impressively, the extensive testing and feedback from "early adopting" universities prior to agency-wide mandatory use in October 2000. These characteristics of FastLane connect with such software engineering concepts as requirement analysis, modes of testing, risk management, and "spiral" software-development models.[19] Similarly, researchers in human-centered computing examine software systems in business and government; the evolution of networking concepts, practices, and institutions; the cognitive and technical dynamics of human-computer interaction; and the effect of varied institutional goals on computing hardware, software, systems, and practices.[20] Our book's emphasis on active users shaping computing developments resonates strongly with software engineers developing "human-centered" models that also highlight user agency and modes of user interaction and feedback.[21]

All these computing researchers recognize the importance of government agencies responding to and shaping the introduction of new information technologies; as Jonathan Grundin put it, "government [is] . . . the largest consumer of computing."[22] Our analysis of how FastLane designers interacted with diverse users, through an iterative process of design, development, and testing, indicates one path for improving scientific understanding and technical design of computing systems. Our extended case study connects also to recent human-centered computing research examining the process of design, models for learning, modes of communication, and the social, cultural, and ethical contexts of computing.[23]

FastLane may even be a useful way of addressing and appraising Conway's law. In 1968 Melvin E. Conway, a computer programmer, consultant, and publicist published a two-page article in the trade journal *Datamation* with the title "How Do Committees Invent?" At the time, the field of software engineering was just forming, and there was interest in how organizations shape and structure software. As Conway phrased his bold hypothesis, organizations designing an information system will produce a design whose structure mirrors the organization's communication structure. Fred Brooks, in his pioneering *Mythical Man-Month: Essays on Software Engineering* (1975), dubbed the remark "Conway's law."[24] In highlighting the Digital Equipment Corporation and its minicomputer VAX, Tracy Kidder made a similar observation in *Soul of a New Machine*. "Looking into the VAX [mini-computer], [engineer Tom] West felt that he saw a diagram of DEC's corporate organization. He found the VAX 'too complicated,'" Kidder wrote. "He did not like, for instance, the system by which various parts of the machine communicated with each other; for his taste, there was too much protocol involved. The machine expressed DEC's cautious, bureaucratic style."[25] NSF is a decentralized organization consisting of independent-minded scientists and staffers. All the same, the design of FastLane embodied intentional choices in hardware, software, and design. The agency's core values, including sanctity of merit review, reliability, and security—if not maximum speed and efficiency—were designed into the FastLane system from the start.

Since society and institutions can shape the path of computing, as Melvin Conway and Tracy Kidder suggest, we also need to address the question of how computing shapes society and institutions. The notion that "computers change the world" is a staple of popular discourse, but there are many pitfalls in our propensity to grant unlimited agency to computing technology—and, accordingly, to overlook social dynamics and ongoing institutional changes that happen to manifest themselves through technology. Clearly, the bicycle boom of the 1890s and the automobile boom of the early twentieth century were connected to changes in gender mores and suburbanization, respectively, but no one should properly conclude that automobiles "caused" suburban sprawl or that bicycles "caused" women's emancipation from Victorian gender codes. Yet the temptation is eternal. Bicycles, in one overheated turn of phrase, were "an eruption of exuberance like a seismic tremor that shook the economic and social foundations of society and rattled the windows of its moral outlook."[26]

Today, computing similarly seems "an eruption of exuberance." This is so because computing frequently is the means by which social or institutional arrangements are reshuffled or restructured. NSF was a mature agency with established institutional values when its core activities were automated by the FastLane system in the 1990s. Consequential changes in computing technologies (such as the Internet, web browsers, and databases), transforming practices in the sciences (such as interdisciplinary and collaborative research), and insistent pressures from Congress (see chapter 2) all contributed to reshaping the scientific community alongside the design and deployment of FastLane. In our interviews, NSF managers drew attention to the centralizing and standardizing effects of FastLane (as compared with the greater diversity in agency procedures permitted in the paper-based era). We asked hundreds of principal investigators to identify significant impacts of FastLane on their scientific practice; their answers, on the whole, indicated that developments internal to their specific research fields were more important than the software application they used to seek funding. (We return to these evaluative issues in chapter 7.)

How to study a community of 50,000 researchers?

We approached our research with full recognition that FastLane was an unusual computer *system*—hardware, software, networking, policy, and procedures together—and that we might need some unusual research methods. We had some of the needed conceptual tools and techniques at hand. Historians studying computing technologies and institutions, along with historians studying contemporary developments in science and engineering, commonly conduct oral history interviews. For many important topics in contemporary science and technology, there is simply no adequate paper trail to be consulted. Our field's well-developed "research grade" oral histories, based on extensive background research and intensive follow-up questioning, were important tools for documenting and understanding the core designers at NSF, a group of people numbering around two dozen. Document analysis, source criticism, and triangulation among sources and perspectives—the familiar toolkit for all researchers in interpretive disciplines like history—were other readily available tools.

Yet the immense size, scope, and diversity of the population of FastLane users we wished to study was clearly something else. We needed to form an unusually large and diverse sample of *users*, as one of the key design goals of FastLane was to offer the entire national research community effective access to this essential pipeline for research support. The intentions of the core designers were important, but of equal measure was whether their intentions were realized and recognized—or for that matter adapted and modified by the users of the system. Typically, computer designers might be thrilled to get half of all potential users signed up, but with FastLane anything short of a near perfect 100 percent rate of user engagement was entirely unacceptable. Not using FastLane meant essentially dropping out of the NSF research community. Accordingly, we planned extensive interviews with scientists and research administrators across the country, at the largest research universities, such as New York University and Stanford, as well as at smaller and less-well-funded places. We also focused on two institutional categories of particular NSF concern: the numerous EPSCoR states and territories, where research funding was seemingly stuck at low levels, and the diverse group of historically black colleges and universities.[27] Our goal was to create a set of interviews with multiple types of users and to tap into the experiences of a wide variety of institutions across the country.

One aspect of our oral-history-centered research design troubled us. Oral history interviews are typically done with only the most prominent persons in a field, for the simple reason that they are time- and resource-intensive. A serious oral history researcher might do five to ten hours of preparation for conducting an hour-long interview, and there is additional time and expense for transcribing and editing the resulting transcript. We don't always recognize it, but what the world knows about science and engineering, across the board, has a distinct bias toward the most prominent leaders—and an unwitting bias against the rank and file who form the vast majority of the research enterprise. It would be unthinkable to conduct public opinion polls where only the most eminent citizens have a voice. Yet despite the recurrent calls for greater citizen participation and engagement in science and science policy, much of what we know about science and technology comes directly from the perspectives and experiences of the most distinguished public figures, such as Nobel laureates or winners of the mathematicians' Fields Medal or computer scientists' Turing Award, as well as prominent members of the national academies of science and engineering. In our era of corrosive skepticism about science and partisan conflict over pressing technical issues such as climate change, these rifts between scientific practices and public knowledge must be of concern to everyone interested in the vitality of science and engineering.[28]

The National Science Board's biennial *Science and Engineering Indicators* provides an aggregate statistical profile of the research community. There are almost 6 million workers in the science and engineering occupations, according to the latest available figures, as well as 14 million more who have a bachelor's or advanced degree in some science or engineering field—the vast majority of whom will never be interviewed.[29] With these concerns in mind, we resolved to design and develop a software tool to assemble a large and broad sample. With perhaps 100,000 active researchers at any moment and around 2,000 NSF staff, we knew we faced an immense target population. Our small research team spent the better part of three years planning, scheduling, conducting, and transcribing interviews, in all making in-person site visits at 29 universities (see appendix A) and several rounds of interviewing at NSF itself. Fortunately, we were able to assemble a research team that had deep experience with research-grade oral history methods as well as the needed computer savvy to develop an online interview platform, to permit truly massive interviewing that is further described in chapters 3 and 7.

Our "mixed methods" of traditional in-person interviews and innovative online interviews resulted in a unique dataset. We believe it to be far and away the *largest dataset* ever assembled on the origins, development, implementation, and use of computing systems. It contains a thousand hours of detailed comments, assessments, reflections, and suggestions from users of the FastLane system—at all levels of the research enterprise. At each of the 29 universities where we did site visits, from Alaska to Puerto Rico, we interviewed an intentionally wide range of people: provosts or vice presidents for research, deans of colleges, department chairs, laboratory directors, regular researchers, program assistants, and clerical staff. We framed a "quasi experiment" by aiming the online interview platform at these same institutions, including also NSF itself, where we interviewed more than 70 staff members. While there has been extensive interest in "online research," especially by researchers investigating sexual and mental health issues, there is not a consensus on whether the online interviews are always comparable with standard in-person interviews.[30] Our research method resulted in a large and reasonably representative sample of on-line responses as well as assessing the comparability of on-line and in-person interviews (see appendix B).[31]

Interviews with NSF program officers, who are especially intensive users of the internal version of FastLane known as eJacket, as well as with diverse NSF policy, administrative, and support staff provide an invaluable cross-agency perspective on NSF's internal operations as well as its interactions with the nation's diverse research community. That we did 79 interviews with NSF staff is, we believe, without peer. In total, our dataset has over 800 interviewees, evenly balanced with half in-person and half online.[32] Around 80 percent of all interviewees allowed their responses to be available to the public. So while this book draws on the entire dataset of more than 800, public and restricted-access, we have created a public dataset of 643 interviews that is freely accessible to other researchers.[33]

Because our project's research questions focused on the design and use of FastLane, we did not systematically mine our dataset for other research questions and concerns. The hundreds of interviews contain a wealth of data on education, careers, research activities, and collaborative relationships for rank-and-file members of the research community. It is a unique dataset of individual qualitative responses to a variety of questions about research policies and practices; it goes into far greater depth and variety than the aggregated quantitative *Science and Engineering Indicators* that NSF publishes biennially. The dataset also may be mined for people's varied expectations, attitudes, experiences, and perspectives on computing. Our interviewees offered us pointed assessments of FastLane compared with other agencies' grant-management systems. And while we ourselves sought to identify "lessons learned" from the several distinct ways of organizing research administration (chapter 7), further detailed analysis of our data might reveal additional insights into effective research policies and procedures.

2

Origins of E-Government

The impulse to use information technology to speed the flow of data and achieve greater precision in its use stretches back to the dawn of the industrial age, when the flow of information itself—the relentless reams of data issuing from commerce, railways, industry, government, and the mathematical sciences themselves—triggered a wave of invention that resulted in the first recognizable computers. Struggling against the plentiful and persistent errors in astronomical tables made by the human computers of the time, Charles Babbage lamented to his colleague, astronomer and mathematician John Herschel, "I wish to God these calculations had been executed by steam." With Herschel's encouragement, and a sizable grant from the British government, Babbage labored for decades to build mechanical computing engines. The implications of his second-generation Analytical Engine were especially vast, for it was not merely a calculating engine but a computing machine capable of conditional branching and susceptible of programming—conceptual hallmarks of modern computing. At issue, as Babbage acutely perceived, was the pace of science and also its direction and, consequently, the character of modern society. "As soon as an Analytical Engine exists, it will necessarily guide the future course of the science."[1]

There are meaningful comparisons between the computing work of Charles Babbage in the mid-nineteenth century and the launch of NSF's FastLane in October 2000. Each was a state-of-the-art response to the rising tide of information. To begin, in the wake of Babbage's work, many governments and companies around the world consciously remade themselves into data-driven organizations where budgets, revenues, expenditures, payrolls, taxes, and much else became standardized, quantified, and ready to be reduced to the orderly boxes of a computer spreadsheet. Then there was the dizzying number of inventors and visionaries who sought to best Babbage at his own game, initially creating mechanical computing machines and subsequently developing electronic ones, which exploded onto the scene in the 1950s, just as the modern science-based system of governmentality including the creation of the National Science Foundation—was taking form. Modern society's defining fancy for ceaseless change and perpetual innovation, gaining momentum ever since Babbage's time, made NSF's management of research that promised technological innovation and economic growth, not to mention a winning hand at the Cold War, into an urgent national priority.[2]

This chapter explores the origins of NSF's research policies and practices to help illuminate when and where and why NSF's FastLane emerged in the 1990s. Several threads are needed to account for this instance of full-blown computerization in the federal government. Consider the following questions. How did an independent agency of the federal government—the term is a technically exact one and not merely a generic description—develop the means and competence to build (as chapter 1 suggests) a vast nationwide electronic post office? Given several earlier efforts to computerize core activities at NSF, what happened in the 1990s to create a supportive environment for a successful effort? How was the effort conceived and coordinated, and who were the people responsible for its emergence—its conceptualization, planning, funding, and even the system's naming? What were the institutional values that NSF sought to "build into" the FastLane system, and can we uncover lessons for today's ventures in value-laden computer design? Chapter 3 goes on to describe how FastLane was developed, implemented, and managed up to its full implementation in October 2000 while subsequent chapters describe and analyze the subsequent experiences of FastLane by scientists, engineers, and research administrators, as well as the NSF staff members who became its most intensive users.

The founding of the NSF and its information implications

The creation of the National Science Foundation as a pioneering civilian funder of scientific research is already a well-told story from political, legislative, and organizational perspectives, so yet another version might seem unnecessary.[3] All the same, as some readers of this book may be unfamiliar with these events—

well known to NSF insiders and science policy experts—we provide a brief history with special emphasis on the *informational* implications of NSF's origins, political context, and evolving mission. With its creation in 1950, NSF was the recipient of a historically specific set of organizational aspirations and expectations. You can clearly see the differences by contrast with the Tennessee Valley Authority or Public Works Administration, created in the 1930s, as well as with the Environmental Protection Agency, created in the 1970s. These federal agencies employed thousands of scientists and engineers, with the express goal of job creation and economic stabilization in the 1930s or support of environmental regulation in the 1970s. NSF research activities clearly enough can result in creating jobs, supporting the economy, and informing environmental management, but these wider societal and economic benefits are not NSF's principal mission. Canonically, it is, according to National Science Foundation Act of 1950 (Public Law 81-507), as follows: "To promote the progress of science; to advance the national health, prosperity, and welfare; to secure the national defense."[4]

Shaping NSF's founding in 1950 were such varied factors as the influential lessons of mobilizing science during World War II, a pitched debate over science and political accountability, and the insistent imperatives of the Cold War, where such seemingly esoteric facts as the precise contours of the earth's gravitational and magnetic fields might be turned to military or geopolitical advantage.[5] The postwar political environment, as we can now appreciate, inscribed specific informational imperatives on the infant agency.

We remember the year 1950 for the momentous events of the Cold War, including the outbreak of the Korean war, the retreat of Chinese nationalists to Taiwan, and the first steps toward an integrated European Community with the Schuman plan, while at home Senator Joseph McCarthy launched his anticommunist campaign and Howard University political scientist Ralph Bunche won the Nobel Peace Prize, an early spark to the civil rights movement. In May of that year, Senator Harley Kilgore from West Virginia was momentarily in the news when Congress passed and President Harry Truman signed Public Law 81-507 creating what Kilgore had some years earlier named the "National Science Foundation." At its founding, NSF represented a delicate balance between Kilgore's politically responsive plans for science, a legacy of the New Deal's embrace of popular democracy and wariness about big corporations and elite control of public institutions, and the recent wartime experiences that seemed to offer different lessons about science and democracy.[6]

Five politically tumultuous years separated NSF's origin in 1950 from what in 1945 might have seemed the taken-for-granted blueprint for organizing postwar science. The plan was penned by Vannevar Bush, a noted engineer, administrator, and inventor of analog computers before the war, as well as the wartime "czar of research" who organized 6,000 of the nation's top scientists and engineers into a formidable research and development machine. Already in the nation's capital since 1938 as head of the Carnegie Institution of Washington, Bush soon became chairman of the National Advisory Committee for Aeronautics (NACA) and before long also of the newly founded National Defense Research Committee (NDRC) as well as the wartime Office of Scientific Research and Development (OSRD). His office launched the Manhattan Project, which spent $2 billion to design and build the two atomic bombs dropped on Japan and also managed the far-ranging research efforts that gave the Allies high-frequency radar and hundreds of other wartime innovations. Bush was at times a "quick-tempered, brusque, and intolerant" administrator; some of his Massachusetts Institute of Technology (MIT) colleagues even characterized him as "arrogant, autocratic, and ferociously ambitious."[7]

Under Bush's leadership, the wartime research agencies spent immense sums of money with extreme dispatch and precious little oversight. No panels of reviewers, evaluators, and supervisors required months of work to select and authorize a research project. Research work might begin within a week or two of initial contact between the OSRD and a promising team, with details and even contracts adjusted down the line to get the job done with minimal delay. One wartime incident cast a portentous shadow, however. Early on, Bush had turned over the massive atomic bomb project to the U.S. Army's misleadingly named Manhattan Engineering District, which maintained such a tight grip on secrecy that vice president Harry Truman himself first learned about the atom bomb's existence only after he was elevated to the presidency when President Roosevelt died in mid-April 1945. Even as president, Truman was not privy to the particulars of the atom bomb project until a full briefing nearly two weeks later. The Manhattan Project was by then fast approaching the Trinity trial. Truman, the "failed haberdasher" and machine politician from Kansas City, Missouri, came to the conclusion that he had been kept in the dark.

With the able assistance of such scientific leaders as University of California, Berkeley's Ernest Lawrence, MIT's Karl Compton, and Harvard's James B. Conant, Bush had run a tight ship during the war. He had no reason to think that the postwar plans for science would depart in any way from the forceful wartime precedents. Released to the public on 19 July 1945, just three days after the Trinity atomic bomb test in New Mexico, Bush intended his report to the president, *Science: The Endless Frontier*, as a blueprint for organizing science in peacetime, articulating a vision of open-ended basic science conducted by civilian researchers, and as an explicit counterargument to Senator Kilgore's proposals for extensive oversight, democratic accountability, and political control. In effect, Bush argued that scientists could and should control their own future. But to President Truman it smacked of the same political elitism that had kept him entirely uninformed of the atom bomb. And now as president he had substantial power. In August 1947 he vetoed an early NSF bill Congress sent to his desk, which would have vested control in an elite board of science experts, insisting on presidential authority to directly choose (and thus to directly dismiss) NSF's director. In his veto message, Truman wrote, "It would, in effect, vest the determination of vital national policies, the expenditure of large public funds, and the administration of important governmental functions in a group of individuals who would be essentially private citizens. The proposed National Science Foundation [in the 1947 version] would be divorced from control by the people to an extent that implies a distinct lack of faith in democratic processes." Truman pointedly rejected Bush's heady vision of scientists owning the research enterprise from the top down.[8]

Even if Bush's political sensibilities were not aligned with the Truman White House, two specific informational consequences resulted from his wartime science empire. First was the basic mechanism of the research grant or contract, which set up a *specific relationship* between the funding agency and an external researcher. "The agency should promote research through contracts or grants to organizations outside the Federal Government. It should not operate any laboratories of its own," Bush had argued in 1945.[9] It was little remarked at the time, but the extramural research grant was a notable departure from the existing mechanisms used by government or industry or philanthropic foundations in supporting science and engineering. The federal government for decades had its own agency-based laboratories, such as the U.S. Geological Survey, the Bureau of Public Roads, the National Bureau of Standards (NBS, today's NIST), and the National Advisory Committee for Aeronautics (NACA, forerunner to NASA). In the main, with the partial exception of NACA, these government agencies did their research work with in-house laboratories and did not form extensive relationships with external researchers. Among the many in-house federal research activities that might be cited, NBS in 1950 built one of the world's first stored-program computers at its laboratories in downtown Washington.[10] During the war, intentionally declining to expand the internal federal-agency laboratories, Bush had directed most of NDRC and OSRD's research funding to external industrial and university laboratories. Wartime research helped transform his own MIT, where the high-frequency radar effort located at the Radiation Laboratory served as a conduit for $1.5 billion in research spending that would transform the institute into a postwar research powerhouse.

For industry, before and after the war, patents were the principal concern. Generally, industry-university relationships devoted great attention to how patents would be identified and issued, as well as licensed and litigated, while paying less attention to nonpatentable research results. Personal relationships and mutual trust count for a lot in relationships between researchers and industry even today. Mutual trust also informed the research policies of the Carnegie, Rockefeller, and Guggenheim foundations, the leading patrons of research before the war. Foundations tended to pick out leading scientific figures or to invest in block grants to research institutes (such as the Marine Biological Laboratory at Woods Hole, Massachusetts, or the Carnegie Institution of Washington), again with less attention to an individual scientist's line of research.[11] With either the industry or philanthropy model, there was little reason for detailed individual research proposals—exactly the type that became omnipresent at NSF—specifying in advance and in detail what research work was to be done.

The political compromise President Truman mandated between the rival Bush and Kilgore visions for NSF also profoundly shaped the new agency's information needs. Bush, seeking autonomy for research, had argued for "complete independence and freedom for the nature, scope, and methodology of research carried on in the institutions receiving public funds," including an appeal for research grants of "five years duration or longer." Such programmatic independence and long-term horizons were necessary to ensure the character of basic research, he felt, as contrasted with short-term applied research aimed at specific needs or outcomes. But Bush was enough of a political realist to recognize that ultimately Congress was paying the bills. By all accounts, he had provided prompt and satisfying testimony many times to Congress during the war years. In *Science, the Endless Frontier*, he fully acknowledged that the "usual controls of audits, reports, budgeting, and the like, should, of course, apply to the administrative and fiscal operations of the Foundation."[12]

Bush's acceptance of financial accountability, strengthened by Truman's insistence on political accountability, meant that there would be significant and ongoing reporting relationships. Reports flowed from individual researchers to NSF's managers, from the National Science Board to Congress, and from the directly appointed NSF director to the president. (As it turned out, the president appointed and the Senate confirmed all 24 members of the National Science Board, as well as the director of NSF.)[13] Even the suggestion that "such adjustments in procedure as are necessary to meet the special requirements of research," bringing a measure of independence from the "usual controls of audits, reports, budgeting, and the like," implied some informational mechanism to establish the political or military necessities that might justify such research.[14]

Two key aspects of NSF's characteristic mode of research management—research proposals and congressional accountability—took form in this political and administrative environment. (A third aspect, peer review, is addressed below.) Research proposals, as they emerged in the 1950s, would be complex documents that sought to justify specific pieces of research and not merely vouch for the general competence or established reputation of an individual researcher or institute. Even in the expansive age of American science that lasted into the 1970s, each NSF proposal made a case for a particular line of research. Subsequently, all researchers awarded NSF funding faced the burden of justifying, initially to NSF and ultimately to Congress, the results associated with their individual research grants. Research practices at other federal agencies were not quite so all-enveloping. The mission-oriented research agencies connected to the Department of Defense had great latitude, at least until the Mansfield amendment to the 1970 Military Authorization Act (Public Law 91-121) sharply restricted the definition of military-related research. In the wake of the Mansfield amendment, Air Force funding ceased at the University of Illinois's cybernetics-oriented Biological Computer Laboratory, leading to its closure.[15] The National Institutes of Health, possibly benefiting from the perception that its research contributed to the "war on cancer" and other high-profile public health campaigns, also took longer-term perspectives on research. Come what may, the new agency's focus on individual research grants and the necessity for congressional accountability meant that an avalanche of paper—proposals, reviews, assessments, and reports—was soon aimed at and flowing through NSF's Washington offices (see figure 5, p. 92).[16]

Peer review in the "golden age" of American science

Peer review of research proposals, as it emerged as a bedrock practice and one of the core values of NSF, was a means to tackle these institutional imperatives of individuality, excellence, and accountability.[17] In framing the agency's research policy, the National Science Board directed the use of "such advisory groups or consultants as the [NSF] staff handling the proposal deems essential." In rapid order a three-level review of proposals emerged, starting with the program officers (or program managers), who were typically subject specialists chosen for their wide knowledge and solid judgment. Proposals deemed to be within the scope of the specific NSF program were evaluated by the imprecisely defined "advisory groups or consultants" in two distinct ways, each of which emerged in the early 1950s and is clearly recognizable today. Some programs, especially in the physical sciences and engineering fields, relied on external written reviews of proposals by specialist researchers. There was a clear hope that these so-called *ad-hoc written reviews*, by specialist reviewers who were, and remain, anonymous (a close parallel to journal-article reviewing), would be detailed and searching; reviewer comments were not automatically forwarded to researchers until many years later, as related below. A second means of engaging external experts was the *panel review*, an early preference in the biological and social sciences, where a group of researchers assembled in Washington to assess the written reviews and discuss the full set of proposals under review. *Site visits* were an additional evaluative mechanism that program officers sometimes used. All of these means for peer review depended on the research community's willingness to invest substantial time in reviewing, as well as on the program officers' knowledge and judgment in arranging a complex process.

Peer review, by whatever means it was conducted, effectively determined the fate of individual research proposals and accordingly shaped the future of scientific fields. For 25 years, until a storm of public controversies in the mid-1970s, peer review was the routine and accepted means for assessing research proposals at NSF. During these years NSF enjoyed substantial autonomy by remaining out of the limelight. It was always the case, however, that peer review was an advisory mechanism to the program officer, who retained decision-making authority on whether to fund an individual proposal. This arrangement, which might seem to be in tension with the hoped-for autonomy of scientific research, or at least with the research community's ability to manage its own affairs, was an acknowledgement that NSF staff—as employees of the federal government—were the proper locus for decision-making authority on the disbursement of federal funds. Accountability was built in to the proposal-review process inasmuch as all proposal-funding decisions by the program officers were routinely subject to additional layers of oversight by upper-level NSF managers. In a pinch, even the National Science Board might be called on to advise on lines of research that raised complicated policy questions. In fact, a set of fractious and often-hostile congressional hearings in the mid-1970s—focusing attention on NSF's internal practices and its controversial Man as a Course of Study (MACOS) educational initiative—would significantly modify the informational aspects of peer review and indirectly shape the earliest efforts at computerizing proposal processing. This chapter is not the place for a detailed account, but suffice it to say that a low point in the agency's public standing was reached when an Arizona congressman accused NSF of promoting, through the $6 million MACOS elementary school social science initiative, such practices as "bestiality, cannibalism, incest," and worse.[18]

With the MACOS controversy still churning, NSF was swept up in the widespread post-Watergate scrutiny of the federal government. The sharp congressional criticism of MACOS led to drastic cuts in the agency's educational activities. "It was the worst political crisis in NSF history," thought one NSF insider. Not until the passage of ten years would educational activities be rebuilt and reconstructed from a second-tier "office" status "with a skeleton staff and few programs" to the full-fledged Directorate for Science and Engineering Education.[19] This same political climate spawned Senator William Proxmire's Golden Fleece Awards. Proxmire's first such award in 1975 went to NSF for a questionably titled study that the senator labeled as a misbegotten effort to identify why people fall in love. He went on to inflict more weighty charges against the Department of Justice, the Federal Aviation Administration, and other agencies. There is reason to believe that Proxmire, as chair of a crucial Senate appropriations subcommittee, closed down NASA's efforts at space colonization. "It's the best argument yet for chopping NASA's funding to the bone," said the influential senator. "I say not a penny for this nutty fantasy."[20]

The Proxmire-style muckraking heralded a political climate where science made all sorts of unwelcome headlines. Long gone were the years where Bush's wartime-era OSRD spent $500 million with minimal congressional oversight and, indeed, without even bothering to create a publicity office. More consequential for NSF's internal operations than the Proxmire blasts, however, was the comment by some members of Congress that the agency was an "old boys club" where (as one congressional critic alleged during a six-day hearing in July 1975) "program managers rely on trusted friends in the academic community to review their proposals. These friends recommend their friends as reviewers," supposedly setting up a closed community of self-perpetuating insiders.[21] This congressional scrutiny led NSF to examine its internal review and award processes minutely. While these years were painful to agency staff, the scientific community gained invaluable insights into NSF's internal (and normally confidential) peer-review practices through the National Academy of Sciences–sponsored research of Jonathan Cole, Stephen Cole, and their colleagues.[22]

The external attention and internal evaluation, in the end, gave NSF for the most part a clean bill of health. Peer review was opened up after the intense scrutiny of the MACOS project, in which program officers seemingly passed along to their NSF staff superiors the positive reviews but did not always include negative ones. From its beginning NSF policy had stipulated strict confidentiality for proposal reviews, which included the complete anonymity of reviewers as well as *not* forwarding verbatim copies of proposal reviews to individual proposers.[23] (Researchers had received only a program officer's summary.) On a trial basis beginning in 1976, researchers gained the option of requesting verbatim copies of all reviews of their proposal; this practice spread across the agency, so by 1981 all proposers as a matter of course received all reviews, positive and negative. Several other long-lived and influential outcomes from this turmoil included a mandated decision turnaround of nine months for all proposals (1975), later shortened to six months; the creation of an internal Office of Audit and Oversight (1977); and the emergence of a standard-length proposal description (1978).[24] Clearly, the next era of proposal review at NSF would put a high priority on speed of reviews, accuracy, accountability, and a standardized proposal length and format. The stage was set for computerizing NSF.

Automation and computerization at NSF

The appointment of IBM's Erich Bloch as the eighth director of NSF in 1984 sent a number of signals—to the public, to Congress, and not least to the agency itself. It rankled some long-time science watchers that Bloch was not a typical academic scientist. He had studied engineering at Zurich's famed Federal Polytechnic Institute then gained a BSEE degree in 1952 from the University of Buffalo, later part of the State University of New York. Joining IBM directly after graduating—he had no PhD—Bloch climbed the corporate ranks to become a staff vice president, including high-profile stints along the way managing the famed Stretch supercomputer and the circuitry for IBM's landmark mainframe System 360. It irked some scientists that even while Bloch expanded research budgets, his priorities seemed inclined toward directly connecting basic science to economic innovation. With the national debate about U.S. industry's "competitiveness" at full tilt, and recurrent headlines about the Japanese electronics and computer companies gaining ground, the early 1980s were a time of national soul-searching about innovation and technical leadership. Bloch had acquired national visibility in 1981 as the first chairman of the Semiconductor Research Corporation (SRC), an arm of the Semiconductor Industry Association. His work for SRC brought him to Washington, where he earned the confidence of George "Jay" Keyworth, then President Reagan's science adviser, who recommended him when the position at NSF opened up.

Inside NSF itself, Bloch raised some worried eyebrows when he made clear that he intended to serve the full six-year term (and indeed he was the first NSF director to do so since the agency's founding director Alan Waterman served out two terms). When agency staff inquired about his plans for the agency, presuming he was a short-termer, he pointedly answered: " 'I might outlive you. So think about that.' That put me on a different relationship with the Foundation's personnel than what the other directors had to deal with."[25]

From an agency-wide perspective, Bloch was well regarded for creating new programs for engineering research centers, supercomputing centers, science and technology centers, and the means to fund smaller high-risk projects known, for many years, as Small Grants for Exploratory Research. He also consolidated the agency's computing programs into a new directorate for Computer and Information Science and Engineering (CISE). Gordon Bell, another SRC colleague and CISE's first director, specifically recalled the Bloch years as exciting and entrepreneurial ones, "even though every congressman and senator tried to influence the outcome for their constituents."[26] Also during these years NSF expanded its high-speed Internet backbone (see figure 2) and then took a leadership role in the interagency National Research and Education Network (NREN), which evolved into Internet 2. NREN "has captured the attention and enthusiasm of an extraordinarily broad collection of interested parties," thought Internet maven Vint Cerf.[27] It was an exhilarating time at NSF. From the vantage point of FastLane, however, the most notable development of Bloch's tenure from 1984 through 1990 was something more concrete.

Soon after arriving, Bloch installed desktop personal computers throughout the agency. "When I came to NSF . . . I was struck by the fact that, like the rest of government, NSF had not prepared for the 'automation of the office' or preparing the management team to be self-sufficient and not depend on administrative help to type letters and similar things," Block recalled. He first set up a personal computer on his secretary's desk and then moved to install PCs across the entire agency. When his staff told him that the maximum number needed across NSF was somewhere between 50 and 150, he boldly rounded up the order to 1,200 because, he noted, there were 1,000 staff people and "because PCs are useful for other things than replacing typewriters." The resulting shifts in secretarial pools and work assignments, he recalled, "caused an uproar . . . it changed the whole environment." Computerization altered the agency's work. "It needed to be done on a wholesale basis. There was no way of starting with fifty computers one year, and then another fifty the next year and then another fifty. And sure, some people probably couldn't master that transition. That's too bad. An organization [has] to move forward in the world. It was a major change for NSF."[28]

The arrival of personal computers triggered experiments at NSF even before the officially recognized FastLane effort. A group of program officers in the Economics Program took steps to automate their own work processes. They initially wired together an early Atari computer with a television monitor so they could run their custom BASIC programs to assist with the ranking of proposals and to assess the financial implications of funding or not funding different proposals. With IBM personal computers and Lotus 1-2-3 spreadsheets, they devised an integrated process to track proposals from start to finish—initial review to final funding decision—including use of a master-budget spreadsheet. Program manager Daniel Newlon recalled that several influential "rules of thumb," including caps on graduate student stipends and budgetary guidelines on computers, travel, and salary, emerged from these early experiments. Email was another important information tool used in Economics. Their pilot effort to receive proposal reviews via email in 1986 or 1987 overwhelmed the existing NSF server, however. Newlon recalled greater success in using email to assist with electronic declines, to distribute proposal reviews, and to prompt tardy reviewers with pending review requests.[29] NSF staff elsewhere in the agency were also engaging in comparable experiments in "user-driven" innovation (see chapters 6 and 7).

The largest effort at computerizing NSF's grant submission prior to FastLane was a joint project done by computer science researchers at the University of Michigan and Carnegie Mellon University (CMU). The project, called Experimental Research in Electronic Submission, or EXPRES, had one title but it evolved to pursue several distinct goals. When the research project began in 1986, NSF intended EXPRES to be a prototype system that would pave the way to agency-wide electronic proposal submission. To achieve this goal, EXPRES examined several available means to create editable "compound documents," combining text and images, before focusing attention on the emerging office document architecture (ODA) standard, an alternative to the automobile industry's rival EDI standards.[30] It happened that the lead Michigan and CMU researchers—Dan Atkins and Jim Morris, respectively—became intrigued with the collaborative authoring of complex documents. NSF's Connie McLindon, too, "immediately got into the problems with collaboration" between researchers with incompatible UNIX, PC-DOS, and Macintosh platforms, remembers one early participant.[31] At the time, there was great interest in the emerging field of computer-supported cooperative work, such as William Wulf's "collaboratories," John Seely Brown's work at Xerox PARC, and Jonathan Grundin's pioneering efforts in human-computer interaction. In general, EXPRES was aiming at developing an email-centered system capable of directly transmitting compound documents like research proposals (as contrasted with the database-driven system later developed for FastLane).[32]

EXPRES was one of several high-profile efforts in multimedia email. Even before 1980, Internet pioneer Jon Postel had publicized the idea of multimedia email while the UUCP (named for Unix-to-Unix Copy) network developed one of the most widely used systems. Some readers may recall using *uuencode* to create ASCII versions of binary files that could be sent via regular text-only email at the time, while the recipient used *uudecode* to reconstruct the original binary files. By the early 1980s there were nine multimedia email projects at half a dozen leading computing research institutions, including Bolt, Beranek and Newman (BBN), the pioneering networking company that had built the ARPANET hardware. EXPRES struck up a partnership with BBN to expand on the firm's Diamond multimedia email project. Users of Diamond could transmit and receive multimedia documents, such as spreadsheets, but they were required to use the editing tools that BBN had built into the system rather than third-party software such as Lotus 1-2-3. Michigan installed a version of BBN's Diamond multimedia email as its local testbed. EXPRES drew also on Carnegie Mellon's workstation-based Andrew system. Andrew, quite similar to MIT's Apollo network, sought to create a campus-wide information system where students, staff, and faculty might log into any workstation on campus, call up their documents, work on them however they wished, then—at least, within the university—send the compound document to a colleague or collaborator.[33]

The network-specific authoring and editing tools developed by BBN and others allowed for an impressive degree of collaboration *within* a specific environment. But the EXPRES project aimed at something wider: the "effective interchange of processable multi-media documents among diverse systems." Processable meant more than today's read-only PDF; it meant that a document created by one person would be fully editable by a second or third person, regardless of their computer systems. So several researchers might collaborate on writing a proposal then have their sponsored projects office also add needed material before sending the compound document on to NSF. It turned out that the Michigan and Carnegie Mellon campus systems, realistically enough for the immense diversity of computer systems across the country, were not only "developed independently" but also "strikingly different in their underlying implementations."[34]

Even though a simple solution for EXPRES might have been to write a translation program between the two different email implementations, the NSF-funded project aimed at a universal system that might work with all conceivable multimedia documents. With each new document type or multimedia system, there would be the need for numerous translators between it and every other existing system. A hypothetical "sixth" system would need five different translations, one for each of the existing systems; a "seventh" system would require six additional translations; and the combinatory demands got progressively worse. The EXPRES team instead chose to translate all documents from all systems to one intermediary standard, the office document architecture; each new multimedia system, regardless of whether there were two or twenty existing ones to match, would need just one translator. Among the immediate challenges was achieving interoperability—roughly, in today's world, achieving uniform style sheets and consistent document formatting—between CMU's Andrew and Michigan's BBN-derived system.[35] "We wanted anybody from any university to send a proposal in any way they wanted to," recalls Connie McLindon. "We knew that meant being able to process multiple, totally different, compound documents . . . it was very challenging."[36]

McLindon played a key role in computerizing NSF during the Bloch years, shaping the EXPRES research, and ultimately taking the steps that launched FastLane. To agency insiders, she is known as "the mother of the son of Fast-

Lane" for her guiding support of EXPRES. In several ways her advocacy of EXPRES and FastLane was a natural follow-on to her earlier work at the Defense Advanced Research Projects Agency (DARPA). In 1973, after she completed an army-sponsored accelerated master's degree at American University and then worked briefly as a systems analyst, DARPA director Stephen Lukasik hired her to be the agency's MIS (management information systems) director. Lukasik, at the time, was managing ARPANET and trying to launch a major effort in artificial intelligence; DARPA recently had been reorganized as a Department of Defense "field agency."[37] She worked closely with DARPA's famed Information Processing Techniques Office (IPTO), including Larry Roberts and Bob Kahn. "I was thrown into the research world . . . not an easy but a wonderful experience," she remembers. She took up program management for IPTO's projects at the University of Southern California's Information Sciences Institute (ISI), Stanford Research Institute, the Institute for the Future, and others. ARPANET was in an early stage. "Everybody had a huge GE terminal, and we used that terminal to log on to California ISI," she recollects.[38]

McLindon brought her DARPA networking experience to NSF when, in 1980, she became the agency's director of the Division of Information Systems. In the next decade, she became senior administrator in charge of policy and implementation of information and communication technologies, serving in a variety of positions at NSF (as her office was reorganized and repeatedly renamed) until 1996, when she left the federal government to become vice president at the Corporation for National Research Initiatives (CNRI), the think tank and innovation shop founded by Internet pioneer Bob Kahn.[39] There, she worked with Kahn to promote the "handle" system that created permanent addresses to locate web-based documents. McLindon's achievements were twice recognized with the Presidential Rank Distinguished Executive Award, once in 1989 for her DARPA work and again in 1996 for her NSF work. Also in 1996, the same year FastLane was awarded for the National Information Infrastructure award for government systems, she received the NSF Distinguished Service Award. The final section of this chapter describes the information technology systems at NSF when she arrived in 1980 and the steps she took to initiate the FastLane project. Chapter 3 picks up the story of FastLane's design and development beginning in 1994 through to its official mandated agency-wide use in 2000.

Information infrastructure at NSF

When McLindon arrived at NSF in 1980, four years before Erich Bloch, her job was to manage the agency's patchwork of computing systems. Unlike her heady experience with research at DARPA, at NSF "they wanted systems improvement [and] not experimentation." The first item on her agenda was retiring the "old Honeywell mainframe . . . running a unique operating system not supported by anybody," which handled the agency's entire range of computing needs, including budgets and personnel files. "I had to get rid of the Honeywell . . . it was a rough couple of years," she recalls. Initially, three Hewlett-Packard (HP) minicomputers replaced the Honeywell mainframe, even though the shift required an entirely different database system and HP did not fully support TCP/IP, so in the longer term, after Bloch "completely transformed the IT environment at NSF," the agency installed an IBM mainframe and retired the HP minicomputers. Her carefully assembled long-range plan for one in four NSF staffers receiving personal computers got an enthusiastic push when Bloch declared that every NSF staffer should have a PC. At the same time, Bloch ordered the agency's financial management director "to plan to automate all financial processes."[40]

It is difficult to imagine today, but in the early 1980s email had not yet come to NSF. At the time, the computer science program officers had email through the phone-based CSNET project (1981–89), which they had funded with a $5 million grant, but the NSF mainframe computer did not support agency-wide emailing. To move things forward McLindon approached the chair of the National Science Board (NSB), Lewis Branscomb, former director of the National Bureau of Standards and at the time the IBM corporation's chief scientist. She suggested getting several of the NSB members on email, and Branscomb enthusiastically agreed to the experiment. "Then everybody wanted to start talking to the National Science Board. So the real senior people got their Dialcom [commercial email] accounts, the assistant directors and above. Then people below that, the division directors, wanted to talk to the assistant directors so [additional] Dialcom accounts there," recalled one agency insider.[41]

With Branscomb's energetic backing, and in time Bloch's as well, email was extended to more and more NSF staff. NSF bought its first commercial email system from Dialcom, a prominent provider also used by the U.S. Department of Agriculture and the IEEE Computer Society. Then, when the rising cost of the commercial system pinched the budget, NSF experimented with an in-house

UNIX-based system that brought email to all NSF staffers and then eventually adopted another PC-oriented commercial system called cc:Mail.[42] This generation of commercial email did not mean seamless connections. "All of the systems were designed to operate autonomously, with no convenient mechanism to allow users of one system to send electronic mail to users on another system," noted one expert in the field.[43] Indeed, a patch written by NSF's Mike Morris permitted cc:Mail to connect to the Internet-based electronic mail. "Everybody eventually got online," remembers McLindon.[44]

During 1992, in the years between EXPRES and FastLane, McLindon was also in charge of NSF's physical move from its longstanding downtown Washington location at 1800 G Street to a new location in suburban Arlington, Virginia. Generally NSF staff liked the downtown location, just two blocks from the White House, and where everyone had "already figured out their commute." When the word came down from the Clinton White House that the move to Virginia was on, despite anyone's reservations, McLindon on one of her first days on the new job (heading not only the agency's IT but also its administrative Resource Management) was taken downstairs to "stand in front of a GSA moving truck for a picture to make it definite," she remembers. The move went smoothly, as it turned out, and even the new computing infrastructure was successfully installed and up and running. The Arlington location, with "superior" working conditions, also featured "plentiful amenities" including nearby "shopping and eating establishments," according to a GAO report. Concerning the move and its many dislocations, McLindon thought, "Neal Lane was a good director for that kind of thing; he was very soothing, and he'd meet with the staff frequently."[45]

A key player on McLindon's energetic team of information technology innovators was Fred Wendling. Even before managing FastLane's design and development in the 1990s, Wendling played a critical role by informally assessing new technologies in the fast-moving field, such as the commercial email services noted above and the local networking described below. Like McLindon, he was an agency outsider with an unusual aptitude and interest in cutting-edge computing. Wendling came to NSF in 1979 with a master's degree in computer science from Indiana University and programming experience for the noted political science research center headed by Elinor Ostrom (later Nobel laureate), which landed him with an interview and a job offer from NSF's Division of Applied Research. He coaxed agency-wide data on comparative funding rates out of the seemingly impenetrable database running on NSF's Honeywell mainframe. His database work attracted Connie McLindon's attention, not least because getting any sort of useful data out of the Honeywell mainframe was a much tougher problem than putting data into it, and she hired him to work directly for her in the Division of Information Systems.

In 1979, NSF's computer infrastructure still centered on the Honeywell mainframe. Nearly all of McLindon's resources went to the 20 computer operators who manned three shifts, around the clock, "to keep the monster up and running." Several Digital Equipment Corporation workstations for document creation rounded out the agency's computing muscle, such as it was. When McLindon retired the Honeywell monster, Wendling gained first-hand experience in automating the agency's workflow by helping write a new reviewer system for the new HP minicomputers. Generating a template for standard reviewer letters that could be customized by individual programs was an attempt to replace the legions of IBM Selectric typewriters. There were staff concerns about jobs. "What I do most of the day is type those letters," he recalls support staff saying to him as he went the rounds explaining the new system.[46]

It turned out that HP's price tag of $300,000 for the first minicomputer, with attractive discounts for the second and third ones, was somewhat misleading. The company made its money selling great numbers of its proprietary terminals at $1,500 each, something like its latter-day gambit with printer-ink cartridges. The HP terminals were customized with a built-in special chip that permitted the display of highlighted input fields on its screen. The investments of the Bloch era, once again, came to the rescue when Wendling wrote a BASIC-language program to turn the ordinary IBM PCs into ersatz HP terminals. Now instead of choosing between PCs and proprietary terminals, NSF could plow its resources fully into PCs that could be used as terminals. And after a "smart switch" was installed in the computer facility (replacing the desktop 9600-baud modems), NSF staffers when they logged in could select a fast terminal connection to the IBM mainframe, the email system, or the HP minicomputers. Wendling's software was eventually picked up by the Walker, Richer & Quinn company, which, on the strength of its HP 3000 terminal emulation, went on to become one of the twenty largest software companies in the country.[47]

In an age when many homes and schools are blanketed with wireless networks and we increasingly take mobile computing for granted, it's important to recall that wired local networking was a major innovation in the 1980s. While IBM promoted its token ring technology, NSF directly connected to the rival Ethernet community, again on the basis of McLindon and Wendling's forward-looking technology assessment. Conceptually, Ethernet was the creation of Xerox Corporation's famed Palo Alto Research Center, where the personal computer as we know it—graphics-rich screen, icons, mouse, networked, with laser printer—first saw the light of day. Inspired by a radio-based packet switching network in Hawaii, Robert Metcalfe and a Xerox PARC colleague invented "Ethernet" as a means to connect the lab's state-of-the-art Alto computers with its one-of-a-kind laser printers. Metcalfe collected a Harvard PhD for the novel design and went on to found 3Com, where Fred Wendling paid him a visit just as the company was developing the first Ethernet cards for the IBM PC. McLindon gave the go-ahead to buy three of the 3Com cards, each around $2,000, and set up an experimental network back at NSF. To make the physical connections, Wendling bought sets of CB radio antenna connectors from Radio Shack to connect the network's coax cables to the back of a computer. When Metcalfe himself stopped by for a visit he learned the lesson: within a month 3Com was offering an entire package of cable, connectors, and crimping tools to support sales and installations of the expensive Ethernet boards.[48]

All the pieces necessary for the wide-ranging FastLane initiative were built up in this way. One final element that came into place, fortuitously, was NSF's relocation to the new suburban Virginia building. Its older downtown building was wired for the hierarchical structure of telephones. Each office in the downtown building was electrically and physically connected to a central switchboard, a neat metaphor for the hierarchical organization of many bureaucratic structures at the time. The "flat" network that Ethernet brought was not easy to achieve in the older building. "We had to go from a PC to the wiring closet back out to the PC to the wiring closet . . . Sometimes we'd pop up, handle three computers, go back down . . . It was an absolute pain to do the wiring," recalls Wendling.[49] The new building's network wiring was one clear advantage of the new Arlington, Virginia, location that NSF was to occupy for two decades.

In the early 1990s, Connie McLindon examined the electronic elements relating to proposal submission that NSF had been assembling and made the decision to go forward with an experimental "Project X." It brought together the lessons from the early networking, Internet, and collaborative experiments; it was, quite clearly, a follow-on to the EXPRES project. McLindon convened a top-level group from the research and industry communities for a "status review" on EXPRES when it became clear that it would not deliver anything like a prototype for electronic proposal submission. A variant based on the emerging PostScript page description language (a creation of the Adobe Systems company, founded in 1982) also came up short on the "effective interchange of processable multi-media documents among diverse systems." Even the automobile industry's EDI standards did not fit the ticket. Clearly enough, electronic grants management was a tough nut to crack.[50]

McLindon assigned Wendling, as he recalls, "the task of coming up with a solution" to enable the movement of proposals from their creation by researchers through review and approval by university sponsored projects offices and onward to peer review at NSF itself. From EXPRES, he adopted the model of "translation" software that would work across the three most-common platforms: UNIX, PC, and Mac. "NSF would [write the software] and send this off free of charge to all the universities," as Wendling recalls the plan. "We were going to create an infrastructure that would allow people to submit to NSF and then other people at the university or other universities to access that information. We'd write the software for three environments." The model of cross-platform document translation was an attractive one, and with McLindon's tentative approval the project was informally begun. Project X might well have become FastLane—except for one small but significant instance of "serendipity." NSF was, after all, on the edge of the web revolution.[51]

3

Developing a New System

During the six years between FastLane's first appearance as an experimental research project in 1994 and its mandatory agency-wide use in 2000, the commercial Internet was born and then exploded. FastLane grew up in its shadow. Early on there was no commercial Internet, and the World Wide Web that created it wasn't so impressive. In January 1994, there were around 600 web servers in the world, a fraction of the 5,000 or more Gopher servers that used the noncommercial Internet to circulate all manner of multimedia. Gopher was installed by NSF for its online Science and Technology Information System (STIS), which published announcements on upcoming competitions, proposal requirements, target dates, and the like. Beyond NSF, Gopher users found such treasures as thousands of scanned texts in Project Gutenberg, or, infamously, on a server somewhere in the South Pacific, beyond the reach of any government, flamboyant pornography. Documents, images, folders, and search-engine links were readily at hand.[1] What you could not do with Gopher is see one screen that combined text, images, and links. Computer science students at the University of Michigan were told, "I know you're all very involved with Gopher right now, so prepare yourselves: very soon, the Web will take off. This new graphical thing called Mosaic is going to change everything." This "new graphical thing" was the achievement of the NSF-funded National Center for Supercomputing Applications (NCSA) at the University of Illinois. Little more than a year after Mosaic's release to the public, the number of web servers jumped to 10,000 and soon more than doubled to nearly 24,000 by June 1995.[2]

The next years were an Internet bonanza. Among the breakthrough Internet companies founded in 1994 *alone* were Amazon.com, Cognizant, Cyberian Outpost, Digital River, DreamWorks, EarthLink, Geek Squad, Lycos, T-Mobile USA, Yahoo!—and a Mosaic-derived start-up named Netscape Communications. Hundreds of companies followed in their wake, thousands of investors piled into the stock market, and for month after month the valuations of Internet companies went higher and higher. At a certain giddy moment, Priceline, the online auction travel site, was valued at more than the American commercial airline fleet.[3] Unfortunately, after the tech-heavy NASDAQ index peaked at over 5,000 in March 2000, there was no place for tech stocks to go but down. Many high-fliers simply went bankrupt, like Pets.com. Other companies sold for pennies on the dollar, the fate of an overvalued bundle of software companies that toy-giant Mattel had purchased in 1999 for $3.5 billion. Telecom giants WorldCom and MCI created the nation's largest corporate merger in 1997 and then, five years later, amid allegations of accounting fraud, the nation's largest corporate bankruptcy. Yahoo! struggled for years with the aftermath of its now-worthless purchases of GeoCities ($3.6 billion) and Broadcast.com ($5.7 billion). Amazon experienced a vertiginous drop in its stock price from a bubble-inflated peak of $107 down to $7, but in subsequent years it has fared far better than the transitory dot-coms.[4]

FastLane does not make many lists of the dot-com wonders or the bona fide e-commerce successes. All the same, FastLane's conception, design, development, and launch were the product not only of NSF's internal dynamics (as traced in chapter 2) but also of the tremendous enthusiasm that fanned the dot-com bubble. E-commerce guru Patricia Seybold presciently included FastLane in the 16 case studies she assembled for her fast-paced *Customers.com: How to Create a Profitable Business Strategy for the Internet and Beyond*. At first, it might seem odd to see the National Science Foundation profiled alongside Amazon.com, Dell Computer, Dow Jones, and Cisco Systems, but the parallels between e-commerce and e-government are deeper than we usually remember. Seybold selected FastLane to highlight e-commerce's streamlining of customer-oriented business practices and delivering of personalized services, explicitly drawing attention to the NSF community's enthusiasm for web-based services. One of the agency's most impressive accomplishments was in creating "a vibrant sense of community around the FastLane application. People who interact with FastLane in institutions across the country get together in face-to-face meetings and on-line forums to share tips and give advice" about best practices and "how to overcome internal organizational barriers to success."[5] Our extensive interviews with users analyzed in chapters 4, 5, and 6 amply confirm the active cooperation and far-flung community, as well as some good-natured competition between rival universities, such as when UCLA and University of California, Berkeley, each hustled to submit the first proposal on FastLane.

In retrospect, we can appreciate why the pioneering phases of e-commerce and e-government were so similar: the same technologies that NSF assembled in streamlining its core processes, such as personal computers, email, databases, and extensive local networking, were pivotal in creating the e-commerce explosion of the 1990s. Across the country and around the world, the innumerable local networks (or LANs) built with relatively small-scale Ethernets or rival Novell networks meant that there were millions of computers ready to go onto the web-based Internet just as soon as there was a compelling reason to do so. NSF had taken a leading role in large-scale networking with the high-speed NSFNET backbone and significant experiments in local networking with 3Com's pioneering Ethernet (see chapter 2). The list of notable computer developments that NSF funded, shaped, and massively benefited from includes the National Center for Supercomputing Applications at the University of Illinois. NCSA was ground zero for the software application, initially known as Mosaic but ultimately world famous as Netscape, that sparked the web-based commercial Internet (see figure 4, p. 92). Serendipity struck when NSF's forward-looking technology assessment expert Fred Wendling made an opportune visit to NCSA. This was a moment that NSF had been anticipating for years. As Seybold recognized, "If you're in a position to know what the default technical standards are most likely to become, you can place that sort of bet."[6]

Origins of the commercial web

The first web-specific step that led to the e-commerce explosion occurred in December 1991, when Tim Berners-Lee, a British computer scientist working at Europe's CERN high-energy physics laboratory outside Geneva, Switzerland, sent a trial version of his web-browsing software to the Stanford Linear Accelerator Center, a sprawling high-energy physics complex on the edge of Stanford's campus. He had been working on and off for several years, recently in collaboration with the Belgian computer scientist Robert Cailliau, to marry the power of networked computers with the notion of interlinked documents that had exercised the imagination in computing and library-science circles ever since Vannevar Bush's prophetic "As We May Think" appeared in the July 1945 issue of *Atlantic Monthly* magazine.[7] Bush's vision was filled in when Ted Nelson, Douglas Engelbart, and others in the 1960s used the term "hypertext" to name interactive and personal links between documents. The aim was organizing the world's corpus of knowledge.

There were several lines of attack for this grand endeavor. Bill Atkinson's HyperCard computer program in 1987 created one practical means of navigating "cards" and "stacks" within the Macintosh world. Cards could have a mix of text and images, linked together with the means of performing dynamic logical and numerical calculations on the fly. A set of cards constituted a stack, and stacks on different computers could be accessed within a Mac-based local-area network. "HyperCard was very compelling back then," a Berkeley programmer named Pei-Yuan Wei recalled. "I got a HyperCard manual and looked at it and just basically took the concepts and implemented them in X-windows" to create one of the earliest web browsers.[8] Berners-Lee's World Wide Web was a full and complete implementation of the leading ideas about hypertext and networked computing. It consisted of "browsers" running on individual terminals, workstations, or PCs that had Internet access to "servers" where the data constituting a web "page" was labeled in such a way that it was readily and remotely accessible. Software protocols—the now-pervasive HTTP and HTML—allowed browsers and servers to communicate. URLs created a global address space. The Internet tied the pieces together, wherever in the world they might be.

In Berners-Lee and Cailliau's original proposal to CERN, a key topic was the means to send files between servers and browsers. Dynamic "links" did the work. "A link is specified as an ASCII string from which the browser can deduce a suitable method of contacting an appropriate server. When a link is followed, the browser addresses the request for the node to the server. The server therefore has nothing to know about other servers or other webs and can be kept simple," they wrote.[9] Their scheme, developed for the networked environment within CERN, a worldwide network of researchers and laboratories, aimed at building an information space for the far-flung community of high-energy physicists. The WWW's open system of global addressing and common interface for accessing information, however, coupled to the open protocols of the Internet, meant that such a web-based scheme could expand far beyond any single professional community. The means to realize a "universal" information space were at hand.

There were, of course, significant barriers in making connections between the well-funded high-energy physics users and those "other servers or other webs." While their original CERN proposal suggested that simple terminals and low-end personal computers might eventually have the ability to "read" web documents, Berners-Lee and Cailliau imagined that high-end workstations would be required to create the hypertext documents. It turned out that CERN's pioneering web browser was also a web editor, running on an upper-end NeXT workstation. It also happened that the world's first web server was Berners-Lee's own NeXT machine located in CERN's building 31.[10] Obviously, if the nascent web required such expensive workstations, it was not going to be spreading very far or fast.[11] Using code that Berners-Lee posted at CERN, software developers began making web browsers. A batch of them appeared in 1992. These were the work of four college students in Finland who created Erwise, the HyperCard-inspired Berkeley programmer who created Unix-based Viola, and a programmer at the Stanford Linear Accelerator Center (SLAC) who created Postscript-savvy Midas. The Midas browser, with one of the first-ever browser "plug-ins" calling up the page-layout language called Postscript, allowed SLAC researchers to view fully formatted technical papers on their screens. For its part, the CERN team gained some independence from the expensive NeXT workstations by building a Macintosh-based browser named Samba. These early browsers brought a small flood of users to the WWW and increased the number of web servers from Berners-Lee's singular machine in 1991 to around 600 by the end of 1993.[12] The browser that triggered the commercial Internet, however, would come from somewhere else.

It is oddly appropriate that the World Wide Web landed in the United States at Stanford. The north boundary of SLAC's 400 acre-campus, to the west of Stanford University's main campus, is Sand Hill Road. There, along a scant mile or so, you will find low-rise buildings with some very distinguished names—Kleiner Perkins Caufield Byers, Sequoia Capital, Draper Fisher Jurvetson—on a strip of expensive real estate that is home to 30 or so of the country's best-known tech investors. For Sand Hill Road is the epicenter of Silicon Valley's venture capital, and venture capital would have a lot to do with creating a blockbuster commercial product from the University of Illinois's Mosaic web browser and financing the start-ups that subsequently created the e-commerce boom. At 2875 Sand Hill Road is one of today's most highly regarded firms, famous for early investments in Twitter, Skype, Instagram, Pinterest, Facebook, and other lucrative deals. In January 2012 it announced the creation of a superfund of $1.5 billion, large even by Silicon Valley's outsized standards. "Software is the catalyst that will remake entire industries during the next decade. We are single-mindedly focused on partnering with the best innovators pursuing the biggest markets," stated Marc Andreessen, co-founder and general partner of Andreessen Horowitz.[13] How Andreessen went, in the interval of two decades, from creating the web browser Mosaic to becoming a billion-dollar venture capitalist is the stuff of Silicon Valley legend.

Andreessen was a 21-year-old student programmer at the National Center for Supercomputing Applications in December 1992 when he and Eric Bina, a recent computer science master's graduate, began work on what became Mosaic. Tim Berners-Lee's WWW protocol was making the rounds, and NCSA staffers Dave Thompson and Joseph Hardin did a demo with the Viola browser to their software design group. Andreessen and Bina created a browser to be run, like Viola, on Unix machines via the popular X-Windows interface. "The project lead for Mosaic is Joseph Hardin," Andreessen noted in a NCSA Mosaic Technical Summary in 1993. With Bina doing the hands-on programming and Andreessen attending to project management, publicity, and customer support, the two began issuing preliminary releases within a couple months. Another team member, Aleks Totic, worked in parallel on a Macintosh version of Mosaic. The first internal release of Mosaic in June 1993 had an impressive list of features, including support of web, Gopher, anonymous FTP, and Usenet news; full HTML display with a wide selection of image, sound, and video formats; and complete hypertext support through underlined "links" and "in-lined images." The first public beta release that September fixed several bugs, improved scrolling and navigation, and displayed non-ASCII characters on the screen. The first public "1.0" release in November 1993 added enhancements such as password-protected file transfers, while the 2.0 release in January 1994 added support for html-based "forms" for structured data input. "Forms are . . . not quite finished, but definitely well into the useful stage," stated the release note.[14] Mosaic's capabilities with forms and file transfers would catch the attention of the FastLane team.

Graduating from the University of Illinois in 1993, Andreessen made his way to Silicon Valley where he found Jim Clark. Clark, the entrepreneurial brains behind Silicon Graphics, was then exploring interactive television. It was for interactive TV and not the web browser that Clark initially hired Andreessen and the team of Mosaic programmers. NCSA was actively commercializing its original Mosaic browser through its licensing agent, Spyglass, which quickly signed up Amdahl, Digital Equipment, Fujitsu, IBM, NEC, and other tech heavyweights—distributing a total of 10 million copies by the fall of 1994—as well as providing Microsoft with the start for its own Internet Explorer. Soon enough, Clark saw the obvious: the team that had originally designed Mosaic could redesign it as a new browser—with the same features but brand-new code. To make the break with Illinois and NCSA complete, Clark renamed his start-up company Netscape Communications. Netscape Navigator 1.0 was released in December 1994 (the same month Microsoft licensed Spyglass Mosaic) and soon gained close to a 90 percent market share in the emerging browser market. The public offering of Netscape's stock on 9 August 1995 created a media sensation and landed Andreessen on the cover of *Time* magazine. He was 24 years old and worth $58 million.[15]

NSF's direct role in NCSA Mosaic and the web explosion is little known and underappreciated. NSF had handsomely funded the University of Illinois NCSA since creating it and four other national supercomputing centers in 1986. Indeed, Larry Smarr, NCSA's first director, had sold the idea of supercomputing centers to NSF director Erich Bloch, who also bought the plan for an NSFNET Internet backbone to link the centers, as a follow-on to the pioneering ARPANET and (it turned out) a transitional step to the commercial Internet. "NCSA was a major hotbed of activity, especially with NSF," recalled one center staffer. In 1990, NSF renewed its funding for NCSA in the amount of $123 million, and it gave four additional awards to NCSA's Joseph Harding during 1990–94 to work on various computer-assisted means for collaboration.[16] NSF insiders recall a $40,000 supplemental request to directly fund the Mosaic development effort. In 1994, NSF awarded the University of Illinois $3.7 million for a multiyear project specifically to support NCSA's "further development of the Mosaic software and related activities such as workshops, documentation development, and internet standards activities" and an additional $4.7 million award to use Mosaic for a digital library infrastructure.[17]

During these years, NCSA also took a leading role in developing the Common Gateway Interface (CGI), which made it possible to connect back-end databases to front-end browsers, creating the basic form of today's e-commerce and e-government. Prior to CGI, a web server responded to a browser request by sending out an exact *static* copy of the requested webpage. Text, images, and layout could be combined, but the presentation was fixed. CGI made it possible for a web server to create a *dynamic* web page; results from executable code from scripting languages such as Perl (or other compiled programs) were combined with the browser's formatting of text and images. "The database example is a simple idea, but most of the time rather difficult to implement," stated one early account describing CGI. "There really is no limit as to what you can hook up to the Web."[18] NCSA's CGI—soon familiar to web users through the "cgi.bin" files—created the means for a company such as Amazon to do e-commerce. Amazon could present a website combining text about a book, images from Amazon or advertisers, and results from real-time database lookups about availability, current sales ranking, and the like. This same enhanced functionality created the means for NSF's efforts in e-government.

The extensive ties to NCSA gave NSF an early awareness of Mosaic and an inside track in shaping its development. Charles "Chuck" Brownstein in the Computer and Information Science and Engineering directorate may have been NSF's first convert to Mosaic. He snagged the attention of Connie McLindon, and she naturally turned to her chief technology assessment expert, Fred Wendling. Wendling's first visit to Illinois to check out Mosaic took place in mid-October 1993 (see figure 4). Mosaic was already impressively multiplatform with Unix, Windows, and Mac versions at hand, as well as a little-remembered version for Commodore's ill-fated Amiga in the wings. As Wendling recounted it, when he asked about an "input" field "where you'd send information to the other side," that is, to a host computer or server at NSF, the Mosaic team answered they'd done it just the week before. What sealed the deal for Wendling was the answer to his speculative question about file transfers: "We're talking about it," they replied. Wendling quickly saw that Mosaic could be the front-end for proposal submissions, trimming untold hours of programming effort.

Wendling understood that Mosaic had changed the game, and he prepared a glowing report for McLindon. By providing structured forms, multimedia file transfers, and multiplatform support, Mosaic had cracked the combinatorial translation problem that had bedeviled EXPRES. Even better, instead of the library of translation programs needed in the "Project X" scheme, a system running on a Mosaic front end would take form as one single web-based Internet application. With McLindon's backing, Wendling returned to Illinois and pestered the Mosaic team about their browser's capabilities—making it impossible for them to forget NSF's keen interest in forms and file transfers. The idea of security was never far from the NSF team members' minds—many months before e-commerce pioneers pinpointed the security implications of credit-card transactions (as noted below). By January 1994, Mosaic publicly announced support for password-protected file transfers and enhanced "forms" capability. For his part, NCSA's Joseph Hardin, after hearing Wendling's thoughts about using Mosaic for a full-blown mission-critical application such as proposal submission, thought the agency's budding plans to be "too ambitious."[19] There were still many points, large and small, to be addressed.

Naming FastLane

For some months, scarcely six people across NSF knew about the ambitious project to computerize proposal submission. It did not even have a name. McLindon and Wendling made a funding pitch directly to NSF director Neal Lane and his closest advisers sometime late in 1993. Wendling, while generally enthusiastic about Mosaic's unmistakable promise, was also cautious about the prospects. He explained to Lane the inherent uncertainties about any new and untried piece of software; there was no guarantee that the Mosaic team would successfully realize the features NSF needed or, indeed, whether they would even continue programming and development. Then there was the awkward fact that less than one-quarter of all NSF-funded researchers had Internet access at the time; the same held true for sponsored projects staff at universities around the country. A universal proposal submission system—requiring Internet access for all—was obviously some years off.[20]

Any reliance on the trendy World Wide Web's taking off was clearly a high-risk wager. McLindon emphasized that since the web-browser software was entirely new and untried, she wanted the proposal-submission effort to be considered as a research project (like EXPRES) rather than a routine procurement or application programming effort. Lane bought her argument, gave McLindon and Wendling a go-ahead, and approved a first-year budget of $800,000 to fund the necessary work. "It's a research project; it's OK if it fails," he told them. When the meeting was wrapping up, he offered a small but essential piece of advice: "Make sure you get buy-in internally [across NSF] and with the [research] community."[21] Each of these targets, internal and external, required extensive efforts.

Perhaps in an idealized world, a new computer system begins with someone typing up software routines or with a soldering iron and a basket of computer chips, but in modern, well-structured organizations the effort to build a computer typically begins with an internal planning document. Around this time, IBM favored a four-part planning scheme—system architecture, logic design and engineering, design automation, process automation—where only the fourth and final stage resulted in anything recognizable as a physical computer ("how do we actually wire a board; what sequence of wire placements should we make? . . . The output of this phase is the physical computer itself").[22] For MIT's famed Multics time-sharing system, frustrating delays in receiving the necessary software-development tools led to the ARPA-funded research group writing up a 3,000-page preliminary design document, in effect an elaborate virtual computer built entirely on paper. Many computer specialists learned about Multics, after the actual system was already built and in operation, from a book-length digest of the oversize design document.[23]

At NSF, McLindon assigned the task of creating an internal planning document to Tom Weber, a PhD in chemical physics who had been a Member of Technical Staff at the renowned Bell Laboratories for 17 years before coming to NSF as a program officer in 1987. He had advanced through several challenging managerial positions and, during 1992–94, served under McLindon as division director for Information Systems. Weber's committee, with Charles Brownstein, David Garver, and other senior-ranking NSF staff, penned a visionary 10-page document, "NSF Task Force on Electronic Proposal Processing," to make the case internally. The composition of the committee, with upper-tier managers, including people from NSF's financial management and varied research directorates, speaks volumes about its purpose: not so much to prepare a detailed design for a computer system but to create the necessary visibility and internal buy-in that Lane had directed.[24] As a formal planning document, if not an actionable technical blueprint, it provides insight into NSF's aims and aspirations for the still-as-yet unnamed system.

With the benefit of hindsight, there are two eye-catching points in the Weber report, as well as one broad assumption that makes for odd reading today. The early assumption was that the agency's *external* means of interacting with researchers for proposal submission would be altered simultaneously with the agency's *internal* means of handing proposal review, assessment, and decision making; even the name, "electronic jacket," that was chosen for the internal means is eerily prescient. In fact, however, what happened was the external-facing system—that is, what became FastLane—was successfully built during the 1990s, whereas the agency's internal means for handling proposals electronically—known as eJacket—was not fully developed until another decade had passed, as chapter 6 discusses in detail. (The delay was largely a result of funding restrictions: "Our budget office was telling us that if it was . . . new or outwardly focused, we could use program funds," which were relatively ample, but "if it was to maintain an existing system . . . or if it was inwardly focused, then we had to use the administrative funds," which were severely limited.)[25]

The vision behind the Weber report was "interoperability between paper and silicon," where paper and electronic versions of proposals would coexist and be entirely equivalent. The core technical challenge, based on the then-current shift to client-server architecture, was creating an electronic-form "kit" that researchers and universities could use to submit digital versions of proposals to NSF. Alongside the electronic kit was a "reengineered" paper proposal, which would be primed and ready for OCR (optical character recognition) scanning when it arrived at NSF.

The aim at this preliminary stage was "a complete system of paper and electronic support" that "allows the transition over time" (note the important temporal qualification) to greater electronic use. The Weber report went to great lengths to assure its readers there would be *no* mandate such as "you must read [proposals] from the screen." Efficiencies and economies were to be won by creating internal "electronic jackets" that were analogs to the paper-based "jackets" NSF had used for decades to physically collect and organize all reviews and documents related to each submitted proposal. All of this complex reengineering activity was forecast to take place in the five fiscal years from 1995 to 2000, resulting in an agency-wide standard proposal system—at a total cost of just $13 million. Remarkably, in an age of endemic cost overruns and extended computer development schedules, these forecasts about time and money were accurate.

Bolstered by the high-level Weber report, and backed by the $800,000 in funding approved by director Lane, the electronic proposal effort gained momentum. It was time for a name. Connie McLindon, Fred Wendling, and Chuck Brownstein traded 50 or more emails that debated dozens of possibilities. Informally, it was "son of EXPRES," which appropriately recalled the translation-based Project X incarnation. For a time they played a version of Scrabble: "Starts with N and ends with V" (for "NSF" and "visual"). Two somewhat unwieldy names kicked around in February 1994 were Proposal and Award Admin Office Processes, or PAAOF, and Experimental Transmission of Research Accounts, or EXTRA.[26] Finally, facing an imminent meeting with Lane, McLindon directed Wendling to come up with the short list. " 'Fred, we got to come up with a name [right now]'—and so I gave her two of them. And one . . . was FastLane, and she chuckled 'nope.' So . . . after the meeting she goes, 'Fred he went for it . . . FastLane'!" The word from the NSF director was, in this instance, "Connie if it's successful, it was named after me; if it's a failure, I had nothing do with it." Wendling remembered a determined search to make sure that no one else had a trademark or copyright on FastLane.[27] Within a few years, the name came into wider recognition when General Motors named its blog Fastlane, the Department of Transportation used it as a convenient label, and even a television show about undercover police officers in Los Angeles picked up the catchy title.[28]

In the first months of 1994, Wendling worked on a detailed design document that would be FastLane's actual blueprint for the coming years. To launch the development effort, NSF held a competition among four contractors with substantial experience and requisite capability. While the computer-services consultants Cap Gemini and American Management Systems were awarded specialized subcontracts, Compuware landed the prime contract. "A senior person at Compuware realized this could have a lot of impact on the company, and he wanted this project very, very badly. He personally made the proposal in response to our request," recalled Wendling. According to another NSF manager, "Compuware grabbed the ball and just took off running with it. And so they ended getting more and more of the FastLane work."[29] Wendling took the lead in defining the specific technical aims to Compuware. "We went over the projects, and [in effect said] these are the six projects I want to do. Give us a proposal to do it." In April 1994 Compuware's "Initial Description of FastLane Processes and Infrastructure" for the first time outlined a web-based plan for processing proposals electronically; the plan involved designing functional components, specifying software, purchasing hardware, and integrating the pieces into a functioning system.[30] This working document, quite different from the Weber report of just a couple months earlier, is clearly recognizable as a detailed blueprint for FastLane.

McLindon and Wendling's attention early on to "external" buy-in from the research community as well as "internal" buy-in from NSF staff shaped the emergent design. Components of FastLane, such as proposal submission, reviewing, project reports, and others "were selected very carefully to hit different groups in the research community so that there were benefits or something to get them to want to use the system," Wendling recalled. With 30,000 proposals coming in to NSF each year, and a much larger number of proposal reviews, there was a vast community to enroll. A module for automatic viewing of proposal status was aimed externally at researchers who had submitted a proposal and internally at the NSF program managers who might save countless hours previously spent responding to researchers' requests for updates. FastLane's capability for cash advances was aimed at the financial officers and sponsored projects offices at research institutions and not so much at researchers themselves. Visibility for award actions was included "for the general public to look at."

Wendling maintained, "We were trying to cover everybody because we knew an important part of this was building the infrastructure of the research community."[31] Another team member remembered, "The key buy-in was with the people in the building [NSF staff] and the PIs. So this was done, I think, very strategically, that FastLane was built one module at a time."[32]

Today's users of FastLane will instantly recognize in this 1994 report many of the system's present-day characteristics and components. After logging in with a password, users would upload their proposal's documents to a server located at NSF—a key characteristic of FastLane, and quite unlike similar proposal submission platforms such as Grants.gov. Users today do not always realize it, but their proposal is *already* at NSF, on a server within the agency, long before their institution might "press a button" and "send" it in. At that time, the proposal was officially submitted and then moved electronically into the regular proposal processing system (specified as "totally electronic," just as in the Weber report). Proposals, at least as this blueprint forecast, would be entirely multimedia, with text, graphics, pictures, and equations. Sometime "in the future" there could even be audio, video, and other media. At this time, multiple word-processing programs were to be supported—the controversial PDF solution was not yet on the radar—with a standard file upload handled by the Mosaic browser and a translation performed at NSF, in an obvious nod to the EXPRES scheme. The April plan also forecast electronic means of conducting proposal reviews with proposals sent out in various word-processor formats.

One feature of the proposed design that real-life FastLane users never saw, despite its being showcased in *Science* magazine, was electronic access to the proposal's status while it was still in review, including the number and dates of the reviews—both solicited and received—as well as the date of the next panel review. Notably, even though some potential problems with server load were possible "if the PI and/or institution checks every day," this component was to be implemented without any log in or password; unusually, "there will not be any security on who obtains the information," as entering the proposal number would be all that was required. Another element with a strong public-access preference was the announcement of award actions, to be made entirely public on "the day the award is made." The brief and sketchy way that the final two components are described suggests that annual and final project reporting was something of an add-on, with details to be filled in later. Several other possible components were left entirely in outline.[33]

Security was a chief concern explicitly identified in the other five components. "It was painful but from the very beginning—security, security, security," observed Wendling, "You do a presentation, talk to anybody and it's 'what the security going to be?' "[34] Wendling's persistence had encouraged the Mosaic team to implement a password-protected file transfer feature. (At the time so-called anonymous FTP, without passwords, was a commonly used method to download text or binary files from a server.) "It was vitally important to the success of FastLane that no one be able to impersonate a researcher or university official," notes author Patricia Seybold.[35] There was even additional NSF research money that focused the software development at Illinois. NSF's Chuck Brownstein funded NCSA's Larry Brandt, in the center's $3.7 million grant (noted above) to do add-ons and enhancements to Mosaic, including encrypted transmissions that improved security.

Not all was sweetness and light in the ongoing interactions with the Mosaic group at Illinois. NSF was not the only potential customer in the minds of the Mosaic programmers. "We've got a problem," recalled Fred Wendling. "One of them that perturbed me was we were doing the budget form and lots of input fields. I don't remember if it was just for one year or if up to five years was on one page, but Mosaic came out with a new version of the software, and [where] we needed like 250 input fields per form, they cut it down to like 140. And I go, 'Wait, we already designed the form, we're ready to go, and you cut down the number of input fields that are allowed.' And they go, 'Well we didn't promise you could always have that many.' And then we go, 'We need it fixed.' They fixed it." With the rising tide of interest in Mosaic, fanned in substantial measure with interagency funding organized by Larry Brandt, the NSF program manager for supercomputer centers, the flood of visitors and never-ending requests for updates and information quickly overwhelmed the Illinois research lab.[36] The situation at NCSA calmed down after Marc Andreessen and the Mosaic programmers went off to Silicon Valley to make their millions and Spyglass took over the licensing of the original software. Wendling remembered talking later to Joseph Hardin and asking, " 'What are you going to do now [that] Mosaic is out of there?' . . . and he goes, 'Thank god I can go back to my research.' "[37]

Clearly enough, FastLane was present at the creation of the Mosaic web browser and the takeoff of the World Wide Web; it was also at the forefront of emerging models of e-commerce. While some early models of e-government focused narrowly on creating a one-way flow of information from a government agency to citizens, something like a newspaper or other mass-media publication, FastLane from the start embraced interactive flows of information among researchers, reviewers, university administrators, financial staff, and the government. Its use of secure, interactive web-based applications is, in fact, astonishingly similar to the contemporaneous version of Amazon.com, which quickly became a notable e-commerce archetype. In this instance, e-government and e-commerce shared features, similar hardware, and even specific software tools.

Amazon.com took concrete form in November 1994—more than a year after Wendling first saw the Mosaic browser at the University of Illinois and nine months after he and Compuware hashed out the detailed blueprint for FastLane—when Jeff Bezos hired two programmers, drafted his wife, Mackenzie, to do the new company's accounting, and began building an integrated database and web front end in his suburban Seattle garage. Amazon ran its early software on Sun Microsystems workstations while FastLane employed a Sun Solaris server; both were coupled initially to databases built, respectively, with Berkeley's DBM and Sybase's SQL.[38] For its back-office ordering and warehousing operations, Amazon eventually adopted a relational database from Oracle, owing to its established reporting capabilities. Amazon's first two programmers, Shel Kaphan and Paul Barton-Davis, struggled to create a software package for a novel set of business activities: existing software packages for mail-order companies, for instance, simply could not handle Amazon's early business model of taking, charging, and shipping book orders *without* maintaining book inventories. (Amazon's massive warehouses came years later.) Bezos estimated that 85 percent of the software development effort in the early years focused on these back-office issues.

For software tools, Kaphan and Barton-Davis adopted the C programming language, a staple of the Unix world at the time, and the text-scripting language Perl, known affectionately as "the duct tape of the Internet . . . used in all kinds of unexpected ways." Amazon, like NSF and, for years, everyone else as well, relied on Netscape Navigator and a complementary Netscape server product to ensure secure connections over the Internet. One Amazon concern that had no analogy in FastLane was the problem of credit-card payments; for months, Amazon took most of its credit card numbers by telephone and kept them on a separate nonnetworked computer. (Yet even here, the physically isolated "sneaker net" was another mode of early networking recalled by several of our interviewees.)[39] Amazon opened its website for public business on 16 July 1995.[40] The Netscape IPO brought fame and riches to the Mosaic-Netscape team just three weeks later.

The same computing elements that made Amazon.com—the Netscape browser, hardware by Sun Microsystems, the software tools C and Perl, and relational databases—also made FastLane. Both Amazon and FastLane were developing a new model where data for a web page was dynamically loaded from a server-linked database using CGI: like today's Wikipedia, Facebook, much of Google, and many newspaper and magazine webpages, these pages do not "exist" until they are called up by a individual user, permitting dynamic web pages with extensive personalization—including targeted advertisements. One thing is clear: web pages that were "static" and written only with the HTML markup language could not work for database integration. C and Perl were new software tools that the FastLane programmers also found handy, even though the two were not readily compatible. This subtle incompatibility is why FastLane programmer Rich Schneider was present at the creation of a software "fork" that persisted in FastLane for two decades.

In 1994, when Schneider joined Fred Wendling's early team as a Compuware employee, even before the project was named FastLane, no one knew how to design a web platform for e-commerce or e-government that would integrate databases, servers, and web browsers. Schneider came to the project after a spell in the U.S. Army, as well as computer training through a University of California extension program and three year's work experience with the programming services company MAI Basic Four. He came to NSF as a Compuware hire, working under Fred Wendling. "I was writing in Perl as fast as I could, because I loved Perl."[41] His ostensible supervisor at Compuware, Al Gianangelo, led a second team of programmers who were writing FastLane program modules in C. "Al was . . . the database person . . . He was brilliant," recalled one NSF manager.[42] A spirited competition emerged between the two teams. It was an immensely exciting time for the programmers who relished "the whole concept of working on the Internet" and creating an entirely original computing application in a brand-new field. "It was a battle, and because Perl's a lot easier to work in, I beat Al to the punch by a long shot," said Schneider.[43]

Fred Wendling liked the dynamic web pages that Schneider had swiftly put up, and "so the Perl track took off" for the input side. Web pages with input forms were the means of uploading names, addresses, titles, and all manner of unformatted text to the FastLane servers at NSF. Al Gianangelo's C programmers specialized on the output side—presenting the user's screen with the information that had been called up from the FastLane database—and since C programs interface cleanly with the PostScript output forms, "there became a division of labor . . . a very clean split in labor." At first there were a half dozen of the paired input-output forms, then there were twenty, and before long each time there was a revision to NSF's fundamental guiding document, the *Grant Proposal Guide*, with new requirements for the elements of all NSF proposals, an entire squadron of specialized programmers and managers was called in. For each successive version of the guide, there were 45 or 50 individual PostScript forms that needed to be revised, on both the output and the input sides. To this day, FastLane's output side is based on PostScript forms delivering information through an API written in C while its input side is a Java implementation of the original Perl program. Even though more modern programming constructs might improve user-friendliness, "It's become almost prohibitively expensive to pull the train off the track of PostScript forms."[44]

The FastLane team made two controversial decisions to depend on proprietary software. The first, as already mentioned, was the early and consequential connection to Mosaic and Netscape. The input-form boxes that still greet FastLane users were acceptable for plain text, such as names, addresses, and the like. But FastLane clearly needed full-blown file transfers so that research proposals could include formatted text, tables, charts, chemical formulas, mathematical symbols, and other complex graphics. "I remember Fred turning to us and saying, hey, you have to implement this [file upload], and I'm saying, but no browser does that yet. He looked at me and said, don't worry, they will. At the time I thought Fred was insane because as a programmer, I'd never heard of writing [a] production program to something that forward . . . But we know now that Fred was correct . . . I would say within two months of the time that we were putting buttons up that said 'Press here to upload your PDF project summary or project description' sure enough, Netscape was supporting file upload," remembered Schneider.[45]

File transfers were controversial in the mid-1990s because not all browsers supported them, and as a government agency NSF was hesitant to mandate the use of the Netscape browser, the only one that fully did so. The de facto Internet standards published as "request for comments," or RFCs, provided a convenient fig leaf. It worked this way. NSF was perfectly content to cite the Internet community's RFC number 1867 that was published by two Xerox Corporation researchers in November 1995. This RFC, explicitly labeled experimental and with the specific proviso that "this memo does not specify an Internet standard of any kind," nonetheless conveniently specified a general means for "Form-based File Upload in HTML."[46] It just happened that one and only one browser, Netscape, fit the bill. As one university phrased the requirement, "You need a computer with an internet connection and a supported web browser such as Netscape 3.0 or higher because it is the only browser that supports file upload."[47] Not until nearly two years later, in September 1997, did Microsoft's Internet Explorer support file transfers for Windows computers, with a version for Macintosh available early in 1998.[48]

The second piece of proprietary software, the much-discussed Adobe PDF, appeared for the first time in a FastLane announcement on 13 March 1996. "The Proposal Submission function now provides the ability to upload the project description as a portable data format (PDF) file. With this new feature, the Proposal Submission now allows the entire proposal package to be submitted electronically. The fifteen page limit on project descriptions . . . still applies." There was no more controversial single aspect of FastLane than NSF's requirement for Adobe PDF, as chapters 4 and 6 detail. At the time, PDF was distinctly experimental, far from the commonplace it is today. Adobe PDF "was a very nice elegant solution" because it provided a means for the "faithful reproduction of the proposal in multiple environments," said Wendling, although he allowed that "NSF took a lot of heat" for its being a proprietary format. "PDF—it was the ideal solution, but it was not easy to implement at all," recollected Carolyn Miller, who became project manager in charge of FastLane in 1995 after Wendling was promoted to a supervisory position.[49]

Creating an acceptable PDF document for FastLane in the early days took perseverance. Converting documents to PDF was "one of the biggest problems we had in the beginning," recalled NSF training specialist Beverly Sherman. It was a "nightmare for users who refused to buy Adobe, and it was a nightmare for us because we had so many printing problems."[50] One potential alternative to Adobe Acrobat was a free program called Ghostscript,[51] which required an intricate multistep process to create PDFs that, all the same, FastLane might well reject when you tried uploading them. Printing PDFs directly with a generic PDF writer inside the word processor (as is common today) resulted in a seemingly fine PDF document that nonetheless FastLane would *not* accept. Even after purchasing and installing a full-featured version of Adobe Acrobat, at perhaps $200 for an individual copy, users faced a cumbrous process. You first downloaded a preferences file that set up the necessary options for creating the PDF within Acrobat's Distiller. With a word-processing document (e.g., .doc) on the computer screen, the next step was to print out and save an Adobe PostScript version (.ps). You then opened the PostScript file in Adobe Distiller, which finally went to work processing the file, line by line, and creating a PDF document (.pdf) that FastLane would accept. Many users remember that PDFs in the earliest days were often something of a problem. The lucky ones had NSF or institutional funding to purchase the needed Adobe software, and it was common for less-well-resourced universities to provide a centrally available station or service for PDF conversion.

Then as now the Adobe PDF format might easily accommodate many other file types than the text, tables, formulas, and graphics that were common in NSF proposals. "There was a lot of discussion about pictures, video, color," according to Carolyn Miller. A PDF document can be thought of as an envelope that can carry and transport many types of computer-file formats, including formatted text, graphics, audio, video, and even executable code. NSF's restricted use of these extended formats, especially its not permitting audio and video supplements to proposals, is not a result of any technical limitations but rather the policies that NSF put in place responding to congressional mandates in the wake of the MACOS controversy (chapter 2). "NSF policy was really starting to clamp down on the size of proposals," since some researchers were tempted to pad their proposals with supporting documents that had the effect of "getting by the size restrictions."[52] As we discuss in chapter 4, researchers we interviewed who today routinely include video clips or simulation files with their online-published journal articles are sometimes frustrated that they cannot also submit these files with their research proposals.

In the 1990s FastLane also had some determined critics from within the federal government, as several federal agencies were experimenting with electronic proposals and had strongly held ideas about how to proceed. The most determined effort to create an alternative to FastLane came from the National Institutes of Health, which had a lot of clout in Washington and a research budget several times larger than NSF's. A short version of a complex story goes like this: the NIH effort evolved through several iterations to produce the eGrants initiative and, ultimately, the Grants.gov system, which was for years an officially mandated successor to FastLane.[53] Grants.gov has occasioned a great deal of critical comment in the field and several withering reports from the U.S. Government Accountability Office.[54] (User perspectives on Grants.gov and FastLane, based on our extensive interviewing, are discussed in chapters 4 and 5.) The proprietary PDF solution attracted the ire of the Department of Education, which "wanted the [FastLane] project stopped and was very vocal about it," said Wendling. "Because we were [supposedly] going in the wrong direction . . . requiring use of a third-party software."[55] Eventually, more than a year after FastLane became mandated for all proposal submissions in October 2000, researchers were able to convert their proposals to PDF documents online.

Managing user-driven innovation

In many computing development efforts, users are a somewhat abstract presence until late in the development cycle. A typical cycle at IBM historically featured sequential and structured phases of requirements, design, implementation, verification, and maintenance, sometimes labeled the "waterfall" model because it progresses overall in a single direction, just like water falling under the force of gravity. Prospective customers first saw an IBM system toward the end of this years-long process, when important decisions had already been made and only incremental adjustments were still possible. IBM's midrange computers in the 1980s intentionally shifted the company to streamline its development cycles.[56] More recently, when Google was launching its Chrome web browser it did extensive in-house testing and the results seemed to be picture perfect. But evidently no one at Google thought to check up on Microsoft. As soon as regular users began using Chrome, in December 2008, an unexpected problem cropped up: the browser gummed up whenever someone called up Hotmail or Windows Live Mail. "I was able to open Hotmail with a recent build of Chrome. I just couldn't actually read any messages or switch folders," reported one user. Google fixed this problem by spoofing web servers into thinking its Chrome browser was Apple's Safari.[57] One of the signal lessons for software engineering is not merely the attention that users received in the development of FastLane but, more importantly, the specific mechanisms NSF created to solicit and encourage user suggestions.

A core group representing diverse perspectives at NSF took the lead in championing FastLane. Connie McLindon and Fred Wendling were experts in information technology and had extensive contacts in the wider computing networking research community, which is how Ethernet and NCSA Mosaic came to NSF in the first place. Another key member of the core group was Jerry Stuck, who had been active in the precursor EXPRES project; he also had extensive experience with NSF's research activities and connections to the wide-ranging external Federal Demonstration Partnership (FDP). Albert Muhlbauer (a member of the Weber committee), from NSF's finance division, understood how money flowed through NSF, which was ultimately the end point of automating the proposal and reviewing processes. Carolyn Miller was responsible for much of the daily implementation and management from her appointment as FastLane project manager in spring 1995. Dan Hofherr worked as Miller's assistant. In early spring 2000, Craig Robinson (succeeding Miller) took over preparations for the launch of FastLane that fall. Robinson, well into the development, remembered, "I would sit down all the time [with Jerry Stuck] and say, what did you do before? What lessons did you learn from that? And then try to apply it to what we were trying to do at the time." Four other government employees—programmer Aftab Bukhari and computer specialists Ellen Quaintance, Beverly Sherman, and Evelyn Baisey-Thomas—completed the FastLane team, supported by "a large team" from contractor Compuware.[58]

Four streams of feedback from users shaped FastLane during the six years of its development. There were two formal standing committees, an informal internal advisory group, and a wider network of contacts with researchers and university administrators formed through NSF's external training and education. Information of different types, representing diverse perspectives, flowed to the FastLane team from these different groups. No single one dominated. Indeed, the diversity of these layers and perspectives yielded an unusually "complex" environment for developing the hardware, software, systems, and processes that constituted FastLane. From its very beginning FastLane was squarely aimed at users outside the agency.

The review committees formed an alphabet soup. The FastLane Internal Review Committee (FIRCOM), was an upper-level group of NSF staffers who, as NSF director Neal Lane described them in 1996, were the "driving force behind the overall success of the project." Project manager Miller described FIRCOM with an analogy to the National Science Board in that they were "at a very strategic level, trying to direct the directions that FastLane was going . . . they tended to be the people higher up in NSF management." Lane half jokingly referred to the initial 16 universities that made up the FastLane Institutional Coordinators Group as "our guinea pigs" (for testing) that constituted the community of early users. The FastLane Internal Implementation Group (FIIG), with "staff from each of the research directorates," was, according to Lane, "where the FastLane rubber meets the road [bringing] FastLane to the front lines of proposal processing."[59]

FIIG evolved from a program officer advisory group. "If I was working on a specific project, then to get it implemented . . . that's what I would use the FIIG people for," recalled Carolyn Miller.[60] They would help assess the practicalities of implementation and the different people likely to be affected by each step along the way. One issue that involved both FIIG and Jean Feldman's NSF Policy Office was standardizing proposal ratings. Canonically, there were discrete rankings of excellent, very good, good, fair, and poor, but reviewers in the paper days could write in intermediate grades, as in half way between "good" and "very good." Initially FastLane was coded to accommodate "multiple types of ratings" and then later changed to require all reviewers to conform to the standardized and discrete proposal ratings.

There were extensive interactions with the 16 "guinea pig" universities. "I got real chummy with them real fast," recalled Carolyn Miller of the external committee. "The people . . . tended to be people who were very enthusiastic . . . and really wanted for the system to succeed." In creating the 16-member external group, composed mostly of sponsored projects staff and some PI researchers, there was specific attention to "different kinds of institutions," including larger and smaller universities, historically black colleges, community colleges, Native American colleges, and representatives from the EPSCoR states (which are those states that historically have received less than the median share per capita of federal research dollars).[61] In addition to specific feedback, the review committees helped NSF set priorities for the initial six program modules, as well as for the subsequent features. According to Miller, the external review committee was especially influential with the proposal status module. "That was so important to them, so that they . . . were having to continually call NSF and find out what was going on."[62]

The FastLane team also engaged in proactive outreach to the research community. The key venues were the NSF regional conferences, university site visits, and the annual meetings of the professional grants administrators, principally NCURA and SRA, respectively the National Council of University Research Administrators and the Society of Research Administrators. For these meetings FastLane team members arrived early and offered well-attended day-long workshops to spread the word, talk about future plans, and gather feedback from audience members. As one researcher administrator at the University of Hawaii commented, "NSF has always had a presence at any of the professional organizations I've gone to, and they've always been one of the forerunners in . . . taking in comments and questions and responding to it."[63] Many universities made it known that they, too, appreciated personal visits (and since these were externally oriented, NSF's ample program funds could be readily tapped for travel expenses), and NSF gave FastLane "even more funds to go to the EPSCoR states." Soon a half dozen FastLane team members—including Carolyn Miller, Bev Sherman, Dan Hofherr, Florence Rabanal, and Evelyn Baisey-Thomas—were perennially on the road, often accompanied by NSF's policy officer, Jean Feldman. "It was really important to have someone from the policy office with us . . . quite often the questions were not technical, they were policy questions," Miller recalled.[64]

Evelyn Baisey-Thomas made a day-long site visit to the Illinois Institute of Technology in Chicago on Monday, 10 April 2000, and it happens that one of us (Misa) retained three pages of handwritten notes from her presentation and 38 pages of her handouts. With the transition to mandatory FastLane use looming just months away in October, her two-hour presentations were focused, practical, and packed with useful information. Within months, NSF would require the use of FastLane for all proposal preparation and submission; business transactions, including payments and financial reporting; and submission of peer reviews, in addition to the annual and final project reports and post award requests and notifications that had already made the transition to the electronic platform. Overall, one can make an estimate of the relative concerns of the NSF FastLane team at this moment. For her handouts, Baisey-Thomas provided 2 pages on annual reporting, 6 pages on proposal submission, 12 pages on general processes—and no fewer than 20 pages devoted to detailed step-by-step instructions on creating acceptable PDF files. Among the how-to topics were how to use Adobe Acrobat 4 for PC, Mac, and Unix; how to use Acrobat 3 for PC, Mac, and Unix; how to use Acrobat with MS Word; information for Ghostscript users; and how to get PDF files from WordPerfect. There was no way of successfully completing the dozen or so steps needed to create an acceptable PDF for FastLane—absent detailed instructions and oftentimes some determined handholding. Just to download and install the "FastLane.joboptions" preferences file that set up Adobe Distiller required successfully completing six tricky steps.[65] As the 20 pages of PDF instructions clearly indicated, the FastLane training team knew that it was a job to do.

The FastLane project's overriding aim of being "user-centric" percolated down to programmer Rich Schneider. He remembered implementing it in a uniquely personal way by cold calling unsuspecting researchers and sponsored projects staff at universities: "Hey, I'm Rich Schneider and I'm working on FastLane, and I see you're having a problem . . . what do you need?" If they told him, for instance, that a button was unworkable or confusing, he would "come up with a solution, usually within days . . . we're going to bend over backward to make sure this is user-friendly."[66]

NSF program officers, always hard pressed to keep up with their load of proposals, reviews, recommendations, and approvals, were by most accounts somewhat distant from the whirlwind of planning meetings and recurrent consultations. Many program officers prior to FastLane had been in a paper-centered world, and they believed the primary users of computers at NSF were, and should be, the support staff who used word processing to generate letters and other routine correspondence vital to the agency's operations. "There was a fair amount of resistance among the program officers," Carolyn Miller remembered, who thought that their job was making decisions about research proposals and not doing the clerical work of processing proposals. "We had a hard time getting the program officer buy-in . . . [they] didn't have as much influence as they could have had if they had participated more actively."[67]

Lane in his "all hands" speech also cited the contributions of NSF's Division of Information Systems (DIS) since its staffers ultimately were the ones responsible for running the computer infrastructure that FastLane built. FastLane was organized as a special external project during its developmental years and then returned to DIS in 2001, about the time when Andrea Norris came to head DIS from an 11-year career as deputy CIO managing a $2 billion information technology budget at NASA.

Even though coupling a server-side database with a browser front end created the dominant e-commerce platform, as FastLane and Amazon were each doing separately, such a combination was far from an obvious strategy in the 1990s. FastLane was initially implemented using Microsoft SQL servers but soon was moved to a Sybase SQL server recommended by none other than Steve Jobs. Wendling recalled Job's visit to NSF, talking up his NeXT computer, as "the first time I had heard about Sybase . . . and I'll credit [Jobs] with pointing us in the right direction for that one."[68] An added bonus was a convenient interface via Sybase's DB gateway to the main NSF mainframe computers. "All our data could be on the mainframe and we could feed it down to the SQL server or the reverse: it would be on the SQL server and we could migrate it up to the mainframe."[69]

As more and more users came online in the months leading up to the October 2000 target, FastLane experienced crushing levels of use. A rule of thumb was that 80 percent of all proposals come into the agency in the last hour or two before a deadline and, especially early on, "the system couldn't support these kinds of loads."[70] There were two general approaches to take the edge off the severe overloading. One involved trimming down the computational load needed for each proposal, since, at the beginning, the FastLane system conducted up to 20 last-minute checks on each proposal before it was successfully submitted. (Recall that the notion of its "being sent" to NSF is not really accurate since before, during, and after submission a given proposal remains on an NSF server.) Cutting out these last-minute checks was one means to trim the last-minute computer loads. Another approach was simply to spread out the proposal load over hours or days. When proposals came in to the agency in mail carts, it didn't matter quite so much. DIS now found itself at the center of scheduling major program deadlines across the agency so that they did not stack up.

So-called agency-wide proposals, such as the young-investigator CAREER program, caused special problems owing to their unusual numbers. Early on, NSF managers decided that the CAREER program would be among the first to use FastLane, which (as one NSF staffer recalled) "brought the system down. Sixteen hundred proposals came in one day and the system was not ready to handle it." These problems persisted, as Craig Robinson recalled the view from the trenches: the "system had broken down for every CAREER deadline, for every year for three years." A decade later, after the heavy loads of the CAREER program, more than double (3,500 proposals), were spread out over *three* days. "Now I watch it, and four or five proposals come in a minute when CAREER is due. So the change is amazing."[71]

Counting down to 2000

The extensive efforts by the FastLane team to generate both internal and external buy-in, to construct independent modules, and to do an incremental rollout to users allowed NSF to bypass the most pernicious event in software development: the endlessly slipping deadline or, what is often the same thing, the botched launch. Early on, going back to the Weber report, the year 2000 was set for the agency-wide implementation of FastLane, and somewhere along the line 1 October was tapped for its full mandatory use: by then, "NSF and the research community will use FastLane for all major proposal, review, award and other interactions."[72] This deadline became something like the Y2K deadline that arrived on 1 January 2000, where IT managers and administrators simply knew that it would occur and so devoted extraordinary efforts to meeting it successfully. "I believe still to this day, if the [NSF] director [Rita Colwell] had moved that date, as so many wanted to as we were getting close to that deadline, then it would have taken two, three, or four more years to get to where we did, but [her] refusal to move that date was critical to the success; and also to put a commitment and resources toward meeting that date," noted Robinson.[73]

In the event, the launch was a bit anticlimactic since by that time thousands of users were already using FastLane. The very first FastLane transaction was a review submitted on 29 March 1995, and subsequently the percentage of total transactions stepped annually upward from 1 percent (FY1996) to 80 percent (FY2000) on the eve of the October launch.[74] As Jean Feldman recalled, "A huge assessment was the volume of proposals that were coming in electronically [and] voluntarily. That data point right there told us a lot about how the community was using it" even prior to the formal launch.[75]

Craig Robinson was a canny choice as FastLane manager during the months immediately prior to the October 2000 launch. He held a PhD from Penn State in astronomy and astrophysics, and he came to NSF after research posts at the Harvard-Smithsonian Center for Astrophysics and then NASA Marshall Space Flight Center in Huntsville, Alabama. His introduction to FastLane was after driving down to Atlanta, Georgia, to attend a FastLane workshop led by Carolyn Miller. "I found it an interesting system, but frustrating from my first experience with it."[76] He came to NSF in 1998 initially to work on the interactive panel reviewing system, one of the new FastLane functions under development, and then took over from Carolyn Miller in the spring of 2000 when she went on maternity leave.[77]

Robinson's appointment took place just eight or nine months before the looming October 2000 deadline when several "senior managers thought that it was going to fail."[78] But NSF director Rita Colwell, as noted above, stuck to the schedule. Soon after arriving, Robinson, a rank newcomer to the established ways of NSF, teamed up with an experienced senior manager, Robert E. "Bob" Schmitz, formerly the division director of administrative services.[79] The two of them together "was actually an extremely good combination." Schmitz "would slap people on the back and talk with them; he was able to communicate around the building what we were doing, why we were doing it, and keep everyone happy," while for his part Robinson quickly hired two extra staff and made plans to tackle the remaining technical problems.[80]

Robinson brought a practicing researcher's viewpoint directly into the FastLane project, viewing it as "great laboratory where people were willing to experiment and try new things and you have a lot of people in the community interested in it." Robinson especially liked the "creative . . . environment" he experienced at NSF, and he highlighted as a positive lesson the benefit of "consistent programmers and managers" from Compuware and the other external contractors. Even after the formally mandated launch in October 2000, the rising load of proposals coming into NSF, especially for the large cross-directorate CAREER proposal rounds, threatened to overwhelm the system. Some much-needed "year-end" money that bought two new large-capacity Sun servers, installed early in 2001, effectively "increased our capacity by an order of magnitude."[81]

As Robinson—and many others—readily admitted, "for the longest time the biggest problem people had was converting to PDF." FastLane was designed around the use of Acrobat's version 3, and several problems emerged when Adobe released version 4 in 2000. Executives from Adobe were called in, and "we got their attention" when Robinson explained that NSF had more than a million PDF documents, "more PDF files than anyone else in the country," and was experiencing persistent difficulties that might bring down the company's product. The immediate problem was solved through an inelegant workaround of creating a backwards-compatible version 3 PDF document, but that was hardly ideal since the capacity for creating PDFs in the first place still required purchase of Adobe Acrobat (and the multistep processes the program required). "At that point, I had a group of people do nothing but correct these PDF problems because the biggest problem we had was with credibility," Robinson recalled.[82]

A welcome solution to the problems of PDF conversion appeared when a small software company approached Robinson with a web-based product that could transform Microsoft Word documents directly into PDFs. Best of all, they made an attractive offer of $10,000 for site installation and $1,000 in monthly maintenance. Robinson snapped it up. Years later, Robinson allowed, "I never let on to them how much we depended on them for the product . . . I would have gladly paid ten times what they were charging us."[83] Although Adobe might have worried about NSF buying a rival company's product to reduce the need for hundreds of research universities to purchase thousands of copies of its Acrobat software, as it turned out Adobe instead saw the expanding use of PDFs in the scientific community as a step toward making PDF into a widely accepted standard across the computer world. Just possibly, they appreciated not getting called in to solve PDF problems at NSF with such regularity.

In March 2005, a decade after the first FastLane transaction, NSF proclaimed a ten-year anniversary and allowed itself a moment of congratulations. By then, FastLane had taken in over 250,000 proposals and more than a million reviews while sending out total awards to the research community of $24 billion.[84] It became an article of faith in NSF that FastLane permitted a dramatic expansion in proposal numbers and awarded research while essentially unchanged in number of staff. During the five years prior to the October 2000 launch, "the number of proposals coming into the Foundation went from 30,000 to 45,000. It increased 50 percent in five years with no increase in staff," according to Robinson. "Without an electronic system, that wouldn't have been possible."[85]

Yet alongside the achievement of these impressive figures, there were also broader changes in the research community and NSF itself. As one researcher saw it, FastLane "probably was a factor in increasing the efficiency of submitting, and I think that's definitely a factor in terms of getting proposal pressure up."[86] One NSF staffer thought that FastLane "increased the workload of the Program Director. But . . . decreased the workload of the support staff and . . . increased the efficiency of the process a lot."[87] Subsequent chapters examine how these changes came about, and how various user communities beyond and within NSF helped transform, and become transformed by, FastLane.

4

Principal Investigators as Lead Users

National Science Foundation principal investigators are a diverse community of 50,000 or more researchers.[1] While most are scientists and engineers, PIs also come from the social sciences (economists, political scientists, sociologists, psychologists, geographers, archeologists, and anthropologists), humanities (historians and philosophers), and even small businesses.[2] The majority are faculty or academic staff at higher education institutions—from the nation's largest research universities to smaller universities and colleges—with lesser numbers hailing from major nonprofit research organizations, such as the RAND Corporation, MITRE, and the Santa Fe Institute, as well as from science centers, museums, and other institutions. A few are independent scholars.[3] Across the United States, NSF funding is critical to maintaining scientific or engineering laboratories, acquiring necessary equipment, and funding the education of graduate students—the next generation of scientists and engineers. It is also critical to attracting undergraduates to scientific and technical fields, or getting a small business off the ground with cutting-edge science or engineering. NSF leaders, who were keenly aware of this diversity and the differential resources across the research community, aimed to avoid putting any portion of this community at a disadvantage in implementing FastLane. Before their effort, no one had dreamed of designing a far-reaching computer system where so many people might give active feedback and shape computer developments in real time.

Our dataset of 800 interviews is the largest one of its kind that we know of, and our interviews with NSF PIs are also unusually detailed and wide ranging. We interviewed hundreds of PIs in person and hundreds more online. We traveled to 29 higher education institutions—including large research universities, smaller schools, EPSCoR state universities, and historically black colleges and universities—covering the entirety of the United States (including Puerto Rico and the U.S. Virgin Islands). Online interviews (the software platform discussed in chapter 1) with PIs and other staff, mostly from these same schools, created a "quasi-experiment" to evaluate and assess our data collection methods. We paid special attention to whether FastLane might have tilted the playing field for any portion of the research community, and we sought to broadly understand PI perspectives on the introduction and early use of FastLane, its influences on work processes and structures within colleges and universities, and whether FastLane had significant influence on the design and structuring of the research projects they proposed or on their inclinations to submit proposals to NSF in the first place. In short, our focus on PIs as one category of lead users (see chapter 5 on research administrators as a second such category) was to grasp the changing processes, perspectives, and contexts of FastLane.

Paper-based days

Before FastLane all NSF proposals were prepared and submitted in paper form, and they conformed to the agency's evolving policies and practices as outlined in chapters 2 and 3 (see figure 7, p. 94). For decades, researchers created proposals using handwritten drafts and typewriters, though by the 1990s many had changed over to dedicated word-processing machines, computers, and printers. When typewriters were the dominant information-processing tool, some PIs had assistance from department secretaries while many others typed out proposals themselves. The last-minute discovery of errors and the frantic efforts to correct them (using Wite-Out or correcting ribbons, retyping pages, or physically cutting and pasting) added significantly to the struggle and stress of proposal preparation. As one PI at an elite research university recalled, "It was always just a hassle to get those forms done because you had to first fill it out by hand . . . So just the mechanics of getting stuff done was pretty tedious."[4]

Most PIs had to photocopy their completed proposal themselves to produce the required 15 or more copies for NSF, though some had help from department secretaries or sponsored research office (SRO) staff.[5] Before widespread access to photocopiers, mimeograph machines were used.[6] Photographs, symbolic notations, or other graphics were produced by specialized and often time-consuming means, and the resulting sheets were inserted into the proposals by hand. Typically, PIs walked the required approval forms around their campus to get signatures from department chairs, deans, and other administrators, including their sponsored research office's authorization of the project and proposed budget.

Once the proposals were copied and the signature forms were complete, there were the final nerve-wracking hours. PIs themselves often packaged and mailed the box of proposals (commonly using overnight services such as FedEx), though they sometimes had assistance from secretaries, student workers, or SRO staff. Last-minute trips to a FedEx office or even out to the local airport were routine.[7] In a surprising number of instances, PIs even took the extreme measure of booking an airplane trip—from Wisconsin, Minnesota, Puerto Rico, and elsewhere—and hand-delivering their proposal (and sometimes also those of colleagues if they were the designated runner) to NSF headquarters in downtown Washington, DC.[8] On the East Coast, Yale University gave its PIs some valuable extra time to make deadlines by having a university staff member make regular runs to Washington to hand-deliver proposals.[9]

At the other extreme, the University of Hawaii for years had no access to overnight-package delivery and at best could arrange for two-day delivery, placing PIs there at a significant disadvantage to the 48 contiguous U.S. states.[10] PIs at schools in Alaska, Puerto Rico, and the U.S. Virgin Islands faced similar geography-specific hurdles.[11] Meanwhile, PIs fortunate enough to be located at Washington, DC, metro area schools might work on revisions until a scant few hours before a deadline and could routinely hand-deliver their proposals to NSF.[12]

Beginning in the mid-1980s, personal computers became common desktop tools for many faculty and academic staff, as well as for support staff. It is worth pointing out, all the same, that several interviewees reported having *no* office computer well into the 1990s. As one HBCU faculty noted, "The first time I had a computer in my office was at the end of 1999, so having to submit a grant online just didn't make any sense."[13] Even with personal computers, collaborative proposals remained a chore; several PIs working on collaborations recalled using a "sneaker net" in the absence of shared local area networks. "Communal editing . . . meant 'sneaker net.' Those were floppy disks [carried] to each other's computers."[14] Work on collaborative proposals typically involved some navigation of incompatible operating systems (Windows, Macintosh, Unix) and mismatched applications software (Microsoft Word, WordPerfect, Lotus 1-2-3, Microsoft Excel, Quattro Pro). Interchange standards, such as Rich Text Format (RTF), came slowly into common use. Overall, in reflecting on the paper-based days, some PIs expressed that it was, by the same token, "all we knew at the time" and not entirely bad, although "relatively painful."[15] Nearly everyone emphasized the laborious and stressful process of generating, packaging, and mailing paper proposals. A significant number lamented the "waste of paper" and the lost trees.[16] Few PIs were aware that NSF staff printed out all FastLane proposals since the agency's internal system remained paper-based for several years.

Learning about FastLane

News about the coming of FastLane—and that it would soon be the required system for grant submission to NSF—spawned mixed reactions. Many PIs learned of FastLane from their university's sponsored research office when it first became widely available for electronic grant submission in FY1998.[17] Others first heard about FastLane from faculty colleagues or an NSF program officer or at a regional NSF conference or statewide annual EPSCoR meeting.[18] By 1998, a small number of NSF programs required proposal submission using FastLane, while a pilot group of 100 researchers tested an early all-electronic version of the annual reporting system. For most NSF programs, FastLane remained optional for nearly two years, until NSF required FastLane use agency-wide starting on 1 October 2000.

NSF also reached out directly to PIs, as instanced by the Fall 1999 newsletter of NSF's Division of Ocean Sciences. Immediately under a news item about the National Ocean Report reviewed by Vice President Al Gore, Woods Hole Oceanographic Institution director Robert Gagosian, and NSF director Rita Colwell, came the announcement: "FastLane . . . will provide a quick, secure, paperless record and transaction mechanism for all NSF awards, from proposal announcement to award closeout." Even if the "learning curve is steep, . . . computer-based business solutions quickly become essential and routine." With just one-fifth of its proposals at the time arriving via FastLane, Ocean Sciences was working on overcoming technical problems with color printing and collaborative proposals. "This change from paper to electronic submissions will prove to be challenging . . . and we ask for the community's cooperation in making this transition."[19]

However they found out about it, PIs' most common initial reaction to FastLane was trepidation and a fear of the unknown, coupled with a sense that electronic documents and electronic document delivery via the World Wide Web were imminent. One chemist explained his "mixed feelings": "It's certainly the wave of the future, and getting rid of paper, or as much paper as possible, is a very good thing," but "every time there's an automated change, I'm supposed to do more . . . and the standards for what is acceptable are higher at each step of the way. So the workload on the Principal Investigator seems to go up with step in technology improvement."[20] Many PIs anticipated a challenging learning curve and some early glitches, and they hoped learning to navigate the system would not take too much time. Most importantly, PIs expressed keen concern that problems with FastLane might cause them to miss an all-important proposal deadline.

A smaller number of PIs viewed news of FastLane at one of two extremes. One geological scientist from the University of Texas, who was mid-career at the launch of FastLane, completely embraced and celebrated the news of the new system, declaring "thank goodness" and stating that he "welcomed it heartily."[21] A chemist at Jackson State, who later became associate dean for research, was equally enthusiastic: "I'm very much interested in . . . and supportive of technology in the work environment . . . I'm a real proponent of using technology to support our operations."[22] A marine biologist at University of the Virgin Islands, frustrated with the NIH's continued reliance on paper, saw NSF as notably forward-looking. "I really welcomed FastLane. I thought that was a huge step up."[23]

Equally strong opinions, however, existed on the other side fearing this new requirement. In the words of one chemist from New York University, "I stayed paper as long as possible," and when FastLane became mandatory, "I dreaded it." Fearing the possibility of missed deadlines, another PI called it "a thing to be dreaded."[24] While our interview data showed no meaningful correlation between research field, geography, or gender and having strong reservations about the arrival of FastLane, it was the case those with a higher number of years since completion of their PhD, a general proxy for age, tended to have greater than average initial concerns. Many senior researchers, all the same, like this chemist, who was a dedicated Macintosh user who had begun his career as a faculty member at the University of Pennsylvania in the mid-1960s, grew to like the system over time—stating that of all the federal funding agency systems, "it's by far the best."[25]

Since there was a two-year measured phase-in of FastLane, there was notably no agency-wide "big bang." Despite some initial concerns about the new electronic proposal submission system, many PIs chose to submit via FastLane when it was still optional for most NSF research programs. Some researchers did this as enthusiastic early adopters who could not wait to be done with paper proposals. Others recognized it quickly would become mandatory and believed they might as well learn the system sooner rather than later; they did not want to fall behind in mastering the necessary knowledge and skills. Attempting to use FastLane when it was still optional—allowing sufficient time before a deadline—always left the backup possibility of submitting a paper proposal if any serious problems occurred.

The early and optional use of FastLane was encouraged by most, though not all, SROs. Some SRO staff offered group training to their university communities (which relatively few PIs attended, if our interviews are representative), and many more made themselves available to help individual PIs use the system. In FY1998 (ending 30 June 1998) 18 percent of NSF proposals were submitted via FastLane. In FY1999 that proportion increased to 44 percent, and NSF's top management set an agency goal for 60 percent for FY2000, which it easily surpassed with 81 percent.[26] Early in the next fiscal year, by 1 October 2000, FastLane was mandatory. As with the launch of any major IT system, particularly a pioneering web-based system created for a diverse group of external users, the road to full adoption was not always smooth. We now turn to users' experiences and their influence on iterative technical and policy modifications concerning FastLane at the NSF.

Experiencing FastLane

It is far from ideal to capture experiential data from users through interviews conducted a decade to a decade and a half after the launch of FastLane.[27] Nevertheless, the great importance of NSF-sponsored funding, and the overriding requirement of successful submissions by deadlines to gain consideration for such funding, led many PIs to vividly recollect crucial contours of the early FastLane system if not always recalling the minutiae.

When they first used FastLane in the late 1990s or early 2000s, the vast majority of PIs found its screen design and functionality to be basic, well designed, and fairly intuitive. FastLane screens were designed to mimic elements of the paper proposals, which facilitated familiarity and fostered navigation. Many users complained that the number of clicks to get to a desired screen was somewhat cumbersome, but the structure, in the words of a University of Montana archeologist, was "very logical."[28] A chemist from University of South Dakota echoed this sentiment: "Every menu that comes up, it's obvious which thing you need to do . . . cover sheets, or the project summary, or project description, or supplemental documents, so it's easy to know which piece you need to work on . . . You just click that button and it asks you to upload the file . . . It's all very logically arranged."[29]

Despite the overwhelming praise, more than a dozen PIs and a similar number of NSF staff, even among those who were generally positive about the system, highlighted the unforgiving aspects of the *hierarchical* program structure. "If you choose a menu and start going through the hierarchy and you're off base, you can't simply go back," stated an anatomy professor at Howard University.[30] "Sometimes I get lost in the hierarchy," observed one geophysicist.[31] Another PI, who at the time was a faculty member at the University of Nevada, Reno, noted, "The most irritating and counterintuitive [aspect] . . . was that you basically needed to navigate with the built-in buttons on the screen." And if you used your browser's back button, "you could get kicked off."[32] Many users learned this the hard way when they lost a screen of populated data they had input.

From the start, users were warned not to use their browser's "back button" and instead to navigate entirely by using FastLane's on-screen buttons. "Navigation was a real problem. If you hit the back arrow on your browser . . . you'd get kicked out and have to sign back in," one chemist remembered. The most common complaints were regularly conveyed to the FastLane design team by the Division of Information Systems Help Desk. It quickly became apparent that the "back-button" warning was not sufficient—too many users were losing data, losing time, and losing patience. As Fred Wendling recalled, early on the FastLane team contacted the Mosaic group at the University of Illinois National Center for Supercomputing Applications for help in disabling the browser's built-in forward and back buttons. As noted in chapter 2, NCSA was accommodating in some significant measure since NSF provided so much of the center's research funding.[33]

Early FastLane budget forms were notably tedious. It took University of California, Los Angeles, fully 10 slides to introduce faculty and staff to the intricacies of FastLane budgets.[34] The budget screens with their many entries took a long time to fill out, and, as such, they were especially susceptible to "back button" errors or other forms of data loss. "The budget form . . . I always have trouble with that, creating it and uploading it. Somehow, I'm not able to, still. And to things in various fields, I don't know how that part works . . . I have relied on somebody else doing the budget form for me," noted one clearly accomplished mathematics professor.[35] Few PIs ever mastered the Excel-based template that was offered to facilitate budget entry; it remained a mystery for years. At least one PI developed his "own spreadsheet that could print out an acceptable facsimile of the NSF budget form."[36] One clear benefit, however, was the various parts of a proposal, including the budget pages, could be viewed and worked on within the FastLane system (at different times) by multiple people.

Especially on collaborative proposals, FastLane as a commonly accessible platform allowed multiple investigators to view the proposed budget and, most importantly, the core of all proposals, the "Project Description." Though many university departments and centers had local networks by the late 1990s, the FastLane platform facilitated access to a proposal for authorized personnel (using password controls) across the campus or around the world. In this sense, by keeping proposal data on remote servers at NSF, FastLane anticipated what later became a central attribute of today's cloud computing.

While few PIs, even those who unreservedly praised FastLane, indicated that the system (compared to the paper-based days) made them more likely to submit proposals to NSF (the science itself and their careers dictated this), a meaningful fraction did stress that the system made them more likely to submit collaborative proposals. "I submitted a collaborative proposal . . . and it was super easy," stated one science education specialist.[37] It was "much easier to write collaborative and multi-institutional proposals," affirmed a Stanford University experimental physicist; recalling a three-institution collaborative proposal in 2001, "I could not have even considered doing that two or three years earlier."[38] Interdisciplinary research collaboration was a growing trend in many scientific and technical fields at the time NSF launched FastLane and was advocated by NSF and other federal funding agencies. The FastLane platform facilitated collaboration with researchers from different departments, or especially at different universities, and appeared to reinforce and add momentum to this broader trend. FastLane's support for collaborative proposals gave multi-institution research teams greater flexibility in assigning the responsibilities for "lead" and support institutions and, by deemphasizing the "subcontract" award structure in favor of each research team having its own budget and individual award, greater control over their own institution's finances.[39]

Access to the proposal on the FastLane platform also transformed PIs' interactions with SRO staff. Previously, PIs would personally hand to their SRO representative the typewritten or printed-out budget or a floppy disk with the budget on a spreadsheet. Often PIs waited to do this until fairly close to the NSF deadline. In contrast, FastLane allowed the SRO to view budgets as they were under development and review them for correctness (proper salary, accurate fringe rates, and correct application of indirect costs) and compliance (allowable expenditures according to NSF and university policies, as well as adherence to page, font size, and margin requirements). SRO staff and PIs overwhelmingly viewed the early budget access as a benefit. In-person contact between PIs and SRO staff did have a tendency to become "more distant," as one Stanford PI observed.[40] A small number of PIs believed this common shift to electronic communication hindered the close and helpful working relationships with SRO personnel that they had often maintained in the paper-based days. Many more PIs appreciated not having to walk across campus, with a deadline looming, and felt that less in-person contact with SRO staff was not harmful. As a neurogenetics professor at University of Hawaii at Manoa put it, "While the interaction . . . may be less personal in the sense of face-to-face, I don't feel any less benefit."[41]

A significant number of PIs experienced, especially early on, either a substantial slowdown in submitting proposals or crashes of overloaded FastLane servers. Such nicknames as "Slow Lane" and "Fast Pain" cropped up. "It had a lot of bugs . . . a lot of glitches," recalled one geologist.[42] The worst problems occurred at deadlines with the heaviest number of overall submissions, such as the crosscutting Faculty Early Career Development Program (CAREER), which was open to all junior faculty. NSF's computer experts in DIS tried to stay ahead of the curve by adding server capacity and using local time for deadlines both for fairness and to distribute heavy server loads, but this sometimes was not enough. In the early days of FastLane the system loads were difficult to anticipate, and the agency's internal testing could not effectively replicate the experience of a live system. Many PIs waited until deep into the deadline day to upload the final version of their proposal, and thus their SROs were pressing the submit button not merely in the final hour but sometimes within the final minutes before the deadline (typically 5:00 p.m. local time).

As one computer scientist from University of Hawaii at Manoa remembered, "Of course the system would crash . . . then we would have a grace day or something to get it in . . . that was almost always the case, actually, every time something was due—for probably the first year, anyway."[43] "Every action was very slow in response," recalled an astronomer from University of Texas at Austin. "The system would crash on regular occasions. And it just seemed as if you could not reliably recreate something from one session to another when it first got implemented. I recall the NSF program officers at my level being dreadfully unhappy with it."[44] Overall, only a minority of PIs remembered crashes and server problems in the first few years of FastLane. When FastLane system problems prevented PIs from making a deadline, they worked with the FastLane Help Desk or their NSF program officer (or with their SRO who coordinated with NSF staff) and were typically granted an extension. While these system slowdowns and crashes did not unfairly affect PIs, these events frequently caused a great deal of stress for both PIs and SROs. One unhappy PI identified a set of serious problems, including slow responses and lost data, observing, "Fairly quickly PIs began to refuse to work with FastLane, leaving actual data and document entry to grant administrators."[45]

Given the early challenges with the FastLane system, some PIs emphasized that electronic submission should not have been required as early as it was. They believed more of the kinks with FastLane should have been worked out before PIs were required to use it. In the words of a chemist from Purdue University, "I was not pleased with its being mandatory because it didn't work as well when first mandated as I believe it does today."[46]

PDF challenges

By far the most widespread early challenge for the research community in using FastLane was uniform document conversion. Cross-platform compatibility was a long-standing concern, stretching back to the prototype EXPRES project (chapter 2). Even in the prototype Project X that was transformed by the Mosaic browser, the translation of proposal documents was a significant challenge. There were two largely independent elements to this issue—and each could affect a PI. The first was ready access to the specific software needed to convert word-processing documents to a common standard, eventually PDF. The second was proper pagination and graphics rendering after PDF conversion. It was well into FastLane development during March 1996 (as chapter 3 recounts), that NSF announced that Adobe PDF would be used—indeed, practically required—for FastLane. "We didn't even know how to make PDFs back then . . . I remember that being a major sort of frustration in the early days," recalled one University of California, Berkeley, professor.[47]

When NSF first began to require submission by FastLane for certain pilot programs in 1998, the Adobe Acrobat software suite that created PDFs (the company gave away PDF Reader) was not widely held inside or outside the academic community. In the words of the then assistant director of sponsored programs at the University of Nebraska–Lincoln, "We found that you'd have one copy in a department and sometimes no copies of Adobe Acrobat in a department especially in the late 1990s, even . . . into 2000."[48] At UCLA, PIs were sternly warned, "Without these [software] elements," including specific browsers and specific versions of Adobe Acrobat or Distiller, "you will be unable to fully use FastLane" (see figure 6, p. 93).

At that time, Adobe Acrobat was not a trivial expense, an individual boxed copy typically retailing for approximately $150. Many PIs purchased the Adobe Acrobat software suite with personal or department funds; some used an NSF grant. However, to some PIs and faculty administrators, the requirement to purchase specialized commercial software to submit a proposal to NSF unpleasantly rankled. "I remember using FastLane and liking it, except for we had to go and PDF our own things," stated one mathematician dean: "How can you say that something is mandatory and then require that we go and purchase a product in order to submit. We were forced to buy Adobe Acrobat." Efforts by NSF to identify a noncommercial workaround, such as Ghostscript "or one of these other bandit operations[,] . . . the instructions were complicated," did not adequately solve the problem.[49]

The PDF issue also spotlighted differential resources among the country's higher educational institutions. It was common at the largest research universities to simply purchase the Adobe software at the department, college, or university level (and, later, to pay for an inexpensive institution-wide site license). For smaller institutions, or those with more modest sponsored research funds, it was more of a burden—to either the school or the individual researcher. "There was no Acrobat in the department, I checked that," stated one chemist at an EPSCoR-state university. At smaller institutions, a dean's office or SRO staff often became a last resort for converting documents for PIs who lacked access to PDF creation software, and this could add significantly to their workloads. SRO staff everywhere, already straining under the last-minute pressure of verifying institutional requirements and officially submitting proposals, were wary of becoming factories for converting an unending stream of thousands of documents to PDF for PIs each year. "Sponsored projects offices . . . were ending up becoming PDF conversion services," recalled a Berkeley research manager. "The unnecessary traffic at proposal time . . . really bogged sponsored projects offices down."[50]

Two changes in the early 2000s helped with the PDF creation problem. First, the cost of licenses from Adobe declined, and more and more PIs gained access to the software. Over time, PDF became a commonly used standard, and at the very least NSF massively encouraged that development. "I've used PDFs for a long time . . . maybe FastLane got me to adopt it early," thought one paleontologist.[51] The PDF creation problem was finally solved when the FastLane team licensed and adapted an online converter for the platform—making it free to the NSF research community submitting proposals. The converter option also solved the problem of slightly different but incompatible versions of Acrobat—since using different versions of Acrobat (3.0 and 4.0 for instance) in the same proposal caused nightmares for NSF. NSF program officers wanted to assure that proposal reviewers saw the same document that PIs intended at submission, not one garbled with the wrong fonts or haywire graphics.[52] While designating a defined standard was controversial when additional purchases by the research community were required, it became easy when researchers' documents could be converted to PDF on FastLane.

Missing fonts and incompatible Adobe versions were not the only PDF-related problems experienced by PIs, NSF program officers, and proposal reviewers. From early in the National Science Foundation's history, accurate graphical representations of science have been important to proposals, especially in certain scientific and technical fields, including but not limited to the biological sciences, geosciences, atmospheric and space sciences, oceanography, chemical engineering, and materials science. Advancing photographic, printing, software, and other technologies have expanded possibilities—and expectations—for graphical representations in scientific research proposals as well as in scientific scholarly publications. The significant level of concern about graphics rendering with FastLane also anticipated the frequent comments we heard about NSF's needing to more fully embrace the next generation of graphics, including simulations, video, and other representations of science that do not fit neatly onto the static two-dimensional "paper" model that persists through today.

A materials scientist then at the University of Wisconsin captured a common early frustration. "I can remember disasters with doing PDF conversion in FastLane . . . We were . . . pushing up against the deadline . . . And we had a couple figures that no matter what we did they did not seem to PDF. And I have no idea, to this day, if what the reviewer saw had the figures or not."[53] Similarly, a marine and environmental geologist at University of Hawaii at Manoa vividly recollected the problems his oceanographic peers had with PDF conversion on NSF proposals: "They mapped the seafloor, and they loved color contour maps. It's a beautiful way . . . [to] present a synthesis of what they have learned . . . [In the paper days] you remember people having somebody or themselves bring up their fifteen [copies of] beautiful color, high resolution printouts to slide into their proposals. They'd know that every proposal would go out and have this great map in it . . . In the transition to FastLane, I clearly recall significant frustration from colleagues as they moved toward standardizing—or just consistently rendering graphics the way you want to . . . It's just not the plot itself, but it's, 'Can you differentiate colors in the same way?' "[54]

For others, the positioning of graphics rather than subtleties of color was the source of problems. A biologist at NYU recalled that, in the early days of PDF conversion, "there were always problems with images . . . I think it was just trial and error . . . where you put your image—sometimes it would just suddenly disappear or it would skip around the page or whatever. So you do experiments with where on the page you want your image. And finally you got something that was okay."[55] Such shifting of text or images also sometimes pushed a proposal's project description, if it was already near the maximum length in the source document, over the 15 page limit—forcing PIs to modify text or the positioning of graphics.

Early problems with PDF conversion of scientific symbols and mathematical notation were far less common. Many mathematicians, computer scientists, and other scientists and engineers using symbols and notation commonly used TeX (a digital graphic typesetting application developed primarily by software pioneer Donald Knuth and released in 1978) or a derivative known as LaTeX (developed by Leslie Lamport in the early 1980s). While some experienced challenges, such as confusion between floating and embedded images in using these text editors, these problems generally were minor and short-lived.[56] The vast majority of PIs found TeX and LaTeX documents to convert easily and accurately to PDF.

In all, problems with rendering graphics to PDF on FastLane largely have diminished over time—with the most serious challenges being overcome within the first half-decade after the system's introduction. Looking back, these rendering problems stood out for their seriousness (technical challenges to proper rendering are a threat to a fair and equitable merit-review process) and in how they affected PIs in different fields, and categories of research in those fields (the extent to which color and other graphics are critical), to far different degrees. "It is frustrating when you review a proposal and there are graphics that are badly reproduced," noted one physicist.[57] The accurate rendering of graphics was entirely unlike budgets or other proposal elements where knowledgeable support staff within departments, colleges, and SROs often were a ready source of troubleshooting tips. With graphics, since generally nobody else at universities had greater expertise with scientific graphics on computers, it was largely up to the PIs themselves to solve the problem.

Apart from whether their field was graphics dependent, PIs' academic fields appeared to have no identifiable bearing in how they adjusted to FastLane or the level of problems they encountered. The only other exception was that less frequent use of FastLane tended to force PIs to relearn the system anew, particularly if some time had passed since they last used it. While scientists and engineers in some fields can fashion their funding outside of NSF (seeking research funds from NASA, the Department of Defense, Department of Energy, USDA, National Institutes of Health, corporations, or other agencies or organizations), most continued to submit regularly to NSF and also review frequently for NSF. Some social scientists and humanities scholars, however, interacted with FastLane infrequently, and thus periodically "relearning" the system took them some extra time—but that was also true for scientists and engineers who, for whatever reason, also used FastLane less frequently.

HBCUs and EPSCoR schools

Historically black colleges and universities are of special interest to NSF owing to their notable success in graduating African American students with undergraduate degrees in science and engineering and preparing them to earn doctoral degrees. In recent years (2006–2010), fully 36 percent of black doctorates in the biological sciences received their undergraduate degree from an HBCU; similar numbers held for baccalaureate degrees of black doctorates in engineering (32%), physical sciences (44%), and agricultural sciences (52%). (Averaging across all science and engineering fields, the figure is 30.9%, while computer science stands at 30.4%.)[58] NSF's Historically Black Colleges and Universities Undergraduate Program (HBCU-UP) directs support toward the roughly 100 historically black schools across the United States. In FY2009, HBCU-UP was funded at $10 million; and in FY2011, at $12 million.[59] In 2010 the Obama administration proposed consolidating three minority-serving NSF programs—HBCU-UP, the Louis Stokes Alliances for Minority Participation, and the Tribal Colleges and Universities Program—into one omnibus program. At the last moment in budget negotiations, extra funds were found, and these programs retained their independent budgets, identities, and activities. HBCU-UP was recently funded at $14.7 million (FY2014) and $18.7 million (FY2015).[60]

In setting up interviews and visiting four historically black universities—Howard University, Florida Agricultural and Mechanical University (FAMU), Jackson State University, and North Carolina Agricultural and Technical State University (NCA&T)—in the contiguous United States with the highest number of funded NSF proposals in recent decades, several points quickly became apparent.[61] With the exception of Howard University, these schools had fewer NSF-funded projects than many non-HBCUs of relatively equal size, and a higher percentage of funded projects were (undergraduate) education-oriented rather than research-focused. This was also generally true of the one historically black university we visited outside the contiguous United States, the University of the Virgin Islands (UVI). Meanwhile, Howard University was similar to non-HBCUs of comparable size both in terms of number of funded NSF projects and the fact that many of its projects were research-oriented.[62]

Despite the smaller scale of NSF funding and the heavier emphasis on education projects at FAMU, NCA&T, Jackson State, and UVI compared with other schools in our study, PI experiences with FastLane at these universities were similar to our study at large. PIs had both excitement and apprehension in learning FastLane would be required for submission to NSF in October 2000, but they found the system to be logical, intuitive, and overall, user-friendly. "When I first became aware of it . . . I was thrilled," noted one marine biologist.[63]

Fewer institutional resources, however, sometimes led to greater initial challenges for PIs using FastLane at HBCUs. A mathematician at NCA&T, when asked about his initial response to learning about FastLane and that it would become mandatory, replied, "It was inconvenient . . . because I did not have a computer in my office at the time . . . in 1999."[64] Computers were not yet common in faculty offices across the NCA&T campus when FastLane was introduced. "In mathematics a lot of my colleagues did not have computers. I think the speed that was required to use FastLane was a [Intel-based] 386 or something in 2000."[65] To submit his proposal on FastLane, this mathematician used his home computer and a dial-up connection (courtesy of his wife being a student at Duke University).[66] A handful of other PIs we spoke with at NCA&T in other departments did have the needed hardware and networking in their campus office at FastLane's launch, as did most PIs at other HBCUs. Nevertheless, access to necessary technology appeared to be more of a problem at HBCUs than at non-HBCU schools of comparable size and level of graduate programs, including those schools in EPSCoR states.[67]

In 1978, Congress authorized the Experimental Program to Stimulate Competitive Research at NSF in response to concerns of undue concentration of federal scientific research funding in specific states and at certain large research universities.[68] Subsequently NSF, NASA, the Department of Defense, and other federal research agencies have participated in EPSCoR activities. In 1979, there were five designated EPSCoR states.[69] Though the "experimental" aspect of EPSCoR is geared as a "catalyst" to boost research infrastructure and competitiveness (ideally beyond needing to be part of EPSCoR in the future), additional states have been added to EPSCoR rather than any states "graduating" beyond EPSCoR status. There are (as of this writing) 31 EPSCoR jurisdictions—including 28 EPSCoR states, along with the territories of Puerto Rico, Guam, and the U.S. Virgin Islands. EPSCoR activities include a variety of block grants to the states and territories and special funding for individual research proposals.[70]

We conducted research interviews at 11 universities in EPSCoR states and territories. Two schools we visited, Jackson State University and University of the Virgin Islands, are HBCUs in EPSCoR jurisdictions. The EPSCoR initiative sets aside a modest portion of NSF's overall budget—approximately 2 percent in recent years—that is awarded partly for merit-based funding through EPSCoR and partly according to each state's designated scientific and research infrastructure needs.[71] Over the years, EPSCoR has had both strong supporters and critics. Its supporters have amassed substantial evidence that it has done a great deal to advance science and technology, especially in its jurisdictions, through the activities of geographically diversifying funding, and many supporters feel it should get more than 2 percent of the NSF budget. Some critics argue instead that the most pathbreaking science comes from a small number of centers of excellence and that the 2 percent set-aside diverts NSF funding.[72]

EPSCoR intersected with FastLane in several meaningful ways. First, EPSCoR programs were often targeted as pilots for submission via FastLane prior to its requirement for all programs. NSF aimed to be proactive in getting EPSCoR schools up to speed with the networking and software technology and upcoming submission requirements. Some EPSCoR schools, such as the

University of Nebraska, were submitting a high proportion of their proposals via FastLane well before the October 2000 mandatory date.[73]

Second, EPSCoR proposals provided some unique early challenges to using FastLane. Some research proposals getting EPSCoR consideration were of identical form to other NSF proposals—with one or more investigators and a small research staff with graduate-student assistants or postdocs. In fact, proposals going to regular NSF programs that received favorable reviews—but, owing to the press of tight budgets, could not be funded by these programs—optionally went through second-round consideration with EPSCoR. NSF program officers needed to be proactive in seeking such funding.[74] In addition to this alternative funding mode, EPSCoR also considered statewide consortia or other multi-institution grants that involved unusually large numbers of participants and wide-ranging collaborations, and these proposals were unique in form owing to the scores of investigators, typically from different campuses, each needing a two-page CV and the ever-present current and pending documents. These proposals were immense, easily reaching 350 pages.[75] This unusual level of collaboration resulted in highly complex proposals in which many individuals needed appropriate and compatible computing, software, and networking resources. Also, since these were nonstandard proposals, directions on the FastLane system were not always clear.[76]

Aside from the complexity of the proposal, computer networking presented challenges to many EPSCoR higher-education institutions. One computer scientist from the University of Montana, who had served as the university's CIO, led an EPSCoR networking infrastructure project for the state of Montana. While the university had fiber optic networking across campus by the mid-1990s, this PI's project was to bring networking resources to the state's tribal colleges—a category of school NSF recognized and targeted to support similarly to HBCUs, but one that faced acute IT infrastructure challenges.[77] This project was able to meet its goal to bring high-speed connectivity to Salish Kootenai College, whose main campus is located midway between Missoula and Kalispell. Other Montana tribal colleges in the towns of Lame Deer and Crow Agency, both in sparsely settled eastern Montana, were not included in the plan. As this PI put it, "It's tough to get high speed connectivity there . . . We just didn't know how we would do it . . . how much it would cost."[78] However, as this PI stressed, these geographically isolated colleges could still use FastLane with telephone dial-up service. He thought the greatest challenge with these tribal colleges using FastLane and securing NSF funding was not limited technology but instead limited staffing.[79] To a lesser, though still very meaningful degree, the absence or low numbers of support staff hindered many higher education institutions' efforts to maximize sponsored research from NSF and other agencies.

Departmental support for FastLane

Universities varied widely with regard to research administrative infrastructure, organization, and support, even setting aside the notable examples of otherwise successful NSF PIs lacking personal computers or network connection in their offices when FastLane was introduced. HBCU and EPSCoR state schools typically existed on one side of the resource spectrum, contrasted with the largest elite public and private universities at the other end of the spectrum.[80] However, many non-HBCU schools in non-EPSCoR states also faced challenges with resources and sponsored research infrastructure. And some schools in EPSCoR states, such as the University of Utah, have significant sponsored research infrastructure. Chapter 5 addresses upper-level or centrally organized research administration—the work, organization, and user experiences of university-wide research administrators with FastLane. SROs performed standard functions of oversight, compliance, and accounting—all of which were affected to a degree by FastLane. Beyond this, SROs, to a greater or lesser extent, helped to train and assist PIs using FastLane. At most schools there were practical limits to degrees of assistance SROs could provide due to their heavy workloads. As sponsored research has become increasingly important to universities in recent decades, colleges, departments, and centers with greater resources have invested in hiring in-house research specialists—which can very directly affect the level and type of interaction PIs have with FastLane. The rest of this chapter looks at the "local" levels of research administration, often physically located close to faculty PIs, as compared with the "central" level of research administration (covered in chapter 5).

At schools in the top-tier of sponsored research funding—Stanford, Berkeley, NYU, and University of Washington—a substantial number of PIs (up to half) indicated they have little or minimal direct interaction with the FastLane system. Laboratory- or department-level support staff perform much of the budget preparation (after learning from the PI about project equipment needs, planned travel, and how many assistants or postdocs will be hired) and assembling and updating current and pending support. In some cases where PIs develop only the "project description" documents that they pass to department staff to upload, they have virtually no direct interaction with FastLane for proposal preparation.

The Computer Science and Engineering department at the University of Washington is a model of devoting substantial resources to sponsored research support. It also illustrates the type of impact the introduction of FastLane could have on the support staff labor force at universities. Even back in the paper-based days of the early 1990s, this department had multiple staff members helping faculty with proposals, positions that in recent years are titled "grant coordinators." With the introduction of FastLane, some grant coordinators adjusted well and learned the system; others did not. As one current grant coordinator put it, some "couldn't adapt to the new way of doing things. It just eroded their confidence in their skills because they didn't have hands-on, paper . . . it was all on this screen, this box." For one such worker, "switching over to computerized systems was the beginning of her getting so frustrated she quit . . . some people thought FastLane made our job easier. Whereas for others they thought, 'Oh my gosh, I'm a fossil; nobody needs me anymore.'"[81]

Computer science at the University of Washington has had five grant coordinators who support approximately 50 faculty members. These grant coordinators, who each work with about 10 faculty members, are assigned based on faculty research specialties such as artificial intelligence, operating systems, programming languages, databases, robotics, or human-computer interaction. They are there to support nearly all aspects of proposal preparation (with generally heavy involvement in everything but the "project summary" and "project description"). They are often the main interface working with FastLane and other electronic systems and are integral to fostering a department culture of frequent faculty submissions for sponsored research.[82] While five grant administrators in an academic department, or as few as ten faculty to each grant administrator, is somewhat unusual, University of Massachusetts Amherst has long had a similar ratio in its department of computer science.[83]

Grant specialists are common in the academic departments of leading research universities and at top research centers and research institutes that combine multiple departments. An example is the Courant Institute of Mathematical Sciences at NYU, with just over 100 full-time faculty in computer science and mathematics and an additional 6 research faculty. The Courant Institute has five sponsored research administration members—who help faculty develop budgets, edit LaTeX documents to create the Adobe format, and support other parts of proposal preparation.[84]

Grants specialists in designated fields become true experts and allow faculty to spend more time on research and formulating the next grant project. Those with a focus on a small number of agencies typically master the agencies' electronic systems. Although few departments even at elite universities have as many as five grant specialists, all the same the science and engineering departments at most major research universities—such as those at University of Minnesota and Purdue University—tend to each have at least one or two such positions. These specialists are an invaluable local resource to help faculty and research staff with budget preparation and sundry FastLane questions, as well as with other agency- or program-specific requirements. Schools with lesser resources tend not to have grant specialists in individual departments or centers. HBCUs and most schools in EPSCoR states fit in this category, though many non-HBCU smaller universities outside of EPSCoR states also do not have grants specialist positions within their scientific or engineering departments and centers. In other words, HBCU and EPSCoR state schools were not uniquely disadvantaged, but as with other schools outside the top tier research universities, greater portions of the burden to learn and use FastLane fell to PIs.

One category of PI that tended to face even greater challenges from lack of sponsored research infrastructure resources within their organization compared with any university or college is *small business*. This class of PI also had significant hurdles to overcome in the paper days (with detailed requirements for proposals), though challenges were accentuated with the introduction of an electronic system such as FastLane. Not only do these PIs not have grants specialists staff, but they also typically do not have any central support—the equivalent of SROs at universities. In the words of an NSF program officer for the Division of Industrial Innovation and Partnerships, "We have found a number of the small businesses . . . are somewhat handicapped because they don't have the resources and in spite of all the things we try to do to encourage them, I literally send back 20 percent of Phase One proposals that come in."[85] NSF does everything it can to help as this program officer pointed out, especially calling on the "FastLane Help Desk and all the people in DIS."[86] Independent scholars face similar hurdles and also have to rely principally on NSF for support, but they may not have as strong NSF program officer advocates as those in the Division of Industrial Innovation and Partnerships.

PI recommendations for FastLane and other cyberinfrastructures

Countless of our interviewees, often without prompting, compared FastLane to Grants.gov. Grants.gov was designed to be the common electronic proposal and grants management system for the roughly 1,000 distinct research programs of the 26 federal funding agencies in the United States. It was designated as the future federal government standard for grant submission under the Federal Funding Accountability and Transparency Act of 2006. Despite the mandate that all federal agencies would need to switch to use Grants.gov for proposal submissions, difficulties with implementation resulted in the federal government relaxing this requirement and continuing to allow NSF to use FastLane (and other agencies to continue to use their systems, such as NASA's NSPIRES). Grants.gov has become standard for the National Institutes of Health and some other agencies, and as such, many PIs and SRO staff at universities have had experience with it. In recent years, PIs could submit to NSF via either FastLane or Grants.gov, though few have used Grants.gov. NSF received just 1.3 percent of its more than 55,000 proposals via Grants.gov from June 2005 through August 2006.[87] Given the different requirements of federal agencies, designing such a standard system had tremendous challenges from the start. Problems with Grants.gov led many PIs, SRO staff, and department grant administrators we interviewed to praise FastLane as they critiqued Grants.gov. Among PIs' most frequent complaints about Grants.gov were the great complexity and frustrating loss of time resulting from glitches, the confusing design, and substantial uncertainty about whether a proposal had in fact been successfully submitted. Typical terms PIs used to describe Grants.gov included "very confusing," "not easy," and "an incredible pain"—many agreed it took grant submission in the "wrong direction."[88] Some PIs felt they lost days of time trying to use the system.

While praise for FastLane—especially after it overcame the early hurdles of system slowdowns and PDF conversion—was far and away dominant, PIs and department-level grant specialists offered specific design suggestions and general principles for effective design, development, and implementation of cyberinfrastructures. The most frequently cited principles for designing superior systems were (1) frequent and continual feedback from user communities and (2) thoroughly testing the system before rolling it out. These are things that NSF did well from the start, though it tended to get far more feedback from SROs than from PIs and department administrators. The hiccups with server crashes were not the result of failing to test but rather the inability to anticipate loads of a live system at a major deadline. NSF had numerous sessions on Fast-Lane at the primary professional conferences for SROs—the National Council of University Research Administrators (NCURA) and Society of Research Administrators (SRA).[89] And more SRO staff than PIs were on NSF committees assembled early in the project.[90] NSF also sent DIS staff to do outreach at many schools (encouraging visitors from nearby schools to attend)—events that were generally well attended by SRO staff but less so by the NSF PI community.

Some of the most pointed feedback came from department grant coordinators who are the most frequent users of FastLane. With regard to the "check proposal" feature, which is an option for a fully completed proposal just about to be submitted to NSF, one department grant coordinator, who praised FastLane overall, observed, "Putting the 'Delete' button next to the 'Check Proposal'—not good formatting." She identified a quirk with the budget calculation module that had not been addressed for two years, despite her calls to the FastLane Help Desk. For her, this tepid response was unusual, as she generally found the Help Desk very responsive. Speculating as to why the problem remains, she perhaps identified the greatest challenge NSF faces with FastLane: it is now legacy software. "Or maybe they can't [fix it], because when you look at how FastLane is, and they build on top of it . . . I mean I understand . . . you code and then you add more and add more and add more. If there's a problem . . . you have to go figure out where that is, or try to re-create that problem."[91]

Even though online interviews provided responses that generally reinforced the in-person data, the online interviewees tended to be terse, and, overall, a modestly higher percentage were sharply critical of FastLane. Far fewer online interviewees gave a response to our inquiry about "what could be learned from FastLane to better understand and improve the design of cyberinfrastructures." The majority of the online respondents indicated some initial apprehension in learning of FastLane's arrival, though they understood and welcomed the move to a paperless system. They felt that FastLane, with the exception of some early system crashes and PDF conversion challenges, was easy to use, that problems were addressed fairly rapidly, and that, in general, the system worked quite well. The vast majority of the in-person interviews offered rich qualitative data, and we had the benefit of asking follow-up questions in these interviews—to gain context, explore details of what worked, and learn more about what did not and why. The minority of negative responses on the online interviews most often gave terse responses. "Clunky" was the one-word answer from a PI at Berkeley, in response to a question on the initial screen design and functionality of FastLane.[92] We did in-person interviews with 17 faculty and administrators at Berkeley, and we also collected an unusually large number of on-line interviews (more than 100) from California, with prominent Berkeley participation. Overall, the difference in the "depth" of data between the two types of interviews is far more striking than the modest differences in the tone or assessment of the FastLane system.

The area where PIs, both from in-person and online interviews, identified the greatest opportunity for ongoing improvement was in reporting—both annual and final reports for funded NSF projects. Prior to FastLane, PIs created a narrative report on their project and had great leeway in how they organized and presented results. With reporting, a significant number of interviewees longed for the old days—not paper reports per se but more narrative-based (and open-ended) reports, as opposed to modules with specific questions and text boxes that were introduced with FastLane. One common complaint PIs have is not knowing everything that will be asked, so they frequently provide answers for later questions on earlier ones and then have to revise many of their initial answers.[93] Other comments were that the specific questions and text boxes fit some types of projects (research) far better than others (education or human resources development), or that the text boxes do not facilitate a well-formatted report that prints out nicely and can be shared with others.[94]

NSF instituted the new reports structure with the apparent goal of more easily grasping and harvesting data on important results from projects that they could convey to the U.S. Congress and the nation. Whether this has helped to avoid key findings from projects being lost to program officers and NSF's top management is uncertain. Some PIs hinted that a report structure designed so that program officers might spend less time with reports is not necessarily a good thing. A fraction of PIs expressed that the electronic system encourages sequential clicks to steadily progress through NSF staff workflows and that their reports likely get less attention since the introduction of FastLane. This viewpoint was explicitly made by a materials engineering faculty member from Purdue University: "There is nothing more irritating to PIs than to realize that FastLane ultimately *reads* our reports. Every piece that we do, that no one pays attention to, or that we suspect that no one pays attention to . . . we don't know how they use the information . . . Sure, we're helping them fulfill their reporting requirements but it's a community . . . and they're not fulfilling their responsibility of being part of a community."[95]

Multiple PIs expressed the opinion that NSF FastLane was still tied to the paper paradigm, which is limited for certain types of scientific presentations in proposals and reports. Some of the same research fields that suffered challenges with color renderings with PDFs—including biology, geology, and marine and atmospheric sciences—frequently use video simulations in scientific presentations. A marine and environmental studies faculty member at UVI observed, "Some of my collaborators do modeling . . . so being able to incorporate videos of simulations [into FastLane proposals] would be very convenient. Instead, right now, we just have to do screen graphs."[96]

A physicist at University of New Hampshire presented a similar viewpoint: "At some time it might be good to put short video clips in."[97] Currently PIs can add links, but NSF tells reviewers that these are optional to review. The director of the Arctic Region Supercomputing Center at University of Alaska Fairbanks has long sought to include video with NSF proposals. In the "olden days," he stated, he would have sent videotapes; today his modus operandi is including a slide still in the proposal and a URL for reviewers to view or download the relevant video. If the video were embedded into the proposal document (quite feasible since PDF serves as an envelope for multiple document types), it would bypass the problem of the portal possibly being down for maintenance when the review panel meets to evaluate the proposal. Regardless, given the instructions, reviewers opting to follow such links seem unusual. This PI included a slide and link to videos in a recent proposal and then tracked if they were downloaded. "No one did. I looked at the server log . . . no hits. We would've seen them."[98]

Some PIs volunteered suggestions to NSF that had no connection to FastLane, data that are outside the scope of our study. However, a handful of suggestions not directly on topic had meaningful tangential connection to FastLane—with regard to making better use of digital data that an electronic grant submission and management system such as FastLane might facilitate. One PI, who had moved into upper-level research administration at an HBCU, stressed that NSF could be more proactive in identifying and assisting PIs who are doing significant research yet might be still pigeonholed into smaller NSF grants from human resources and educational programs. While this suggestion was not about FastLane per se, this person stressed that as an electronic system FastLane expands possibilities for capturing digital data (project summaries, project descriptions, reports, etc.) for analyzing patterns, making policy, and implementing change. This PI and research administrator believed that more could and should be done with analyzing this electronic data to facilitate HBCU faculty who publish and conduct important scientific research to be mentored by NSF to "graduate" to funding by the research directorates.[99]

Others PIs also believed the digitization of information, facilitated by FastLane, could be compiled, analyzed, and used more effectively to advance the research enterprise. A School of Engineering Education faculty member at Purdue University argued that data could be used to suggest effective collaborations, as well as to ensure quality data through more effective coordination between funded projects. He stressed that surveys are extensively used tools for engineering education research, but oftentimes concurrent sponsored projects sample the same student populations (same campuses—and often the same segment of students at a particular campus), resulting in survey overload. As he put it, "At certain institutions, [students] get surveyed to death, so the survey response rates are plummeting."[100] He believed funding agencies such as NSF could make suggestions on coordinating research to counteract such problems.

While many PIs had constructive suggestions, and a very few had significant critiques, the overwhelming message from PIs was praise for FastLane and its stability and reliability. Many PIs commented that the system has changed relatively little in design—most saw this favorably in terms of not burdening users with learning new design elements and navigation; others emphasized that the system felt quite dated and that it had not evolved to the carefully engineered user-friendly standards of commercial sites like Amazon.com. Even though some felt FastLane could use a major redesign after more than a dozen years, the majority of these individuals still felt it was relatively easy to use and, most importantly, that the system allowed them to effectively manage time and effort. Doing science, after all, is the goal of researchers.

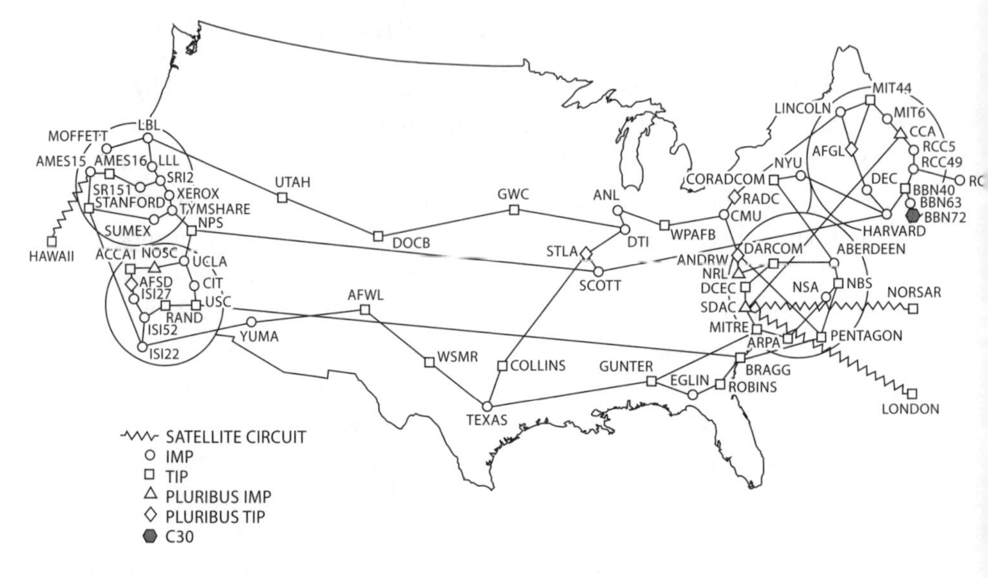

Figure 1. ARPANET map (Oct. 1980). Each named IMP node was linked to one or more local host (mainframe) computers, while a TIP was a standalone terminal connected to ARPANET. *Note:* This map does not show ARPA's experimental satellite connections. Names shown are IMP names, not (necessarily) host names.

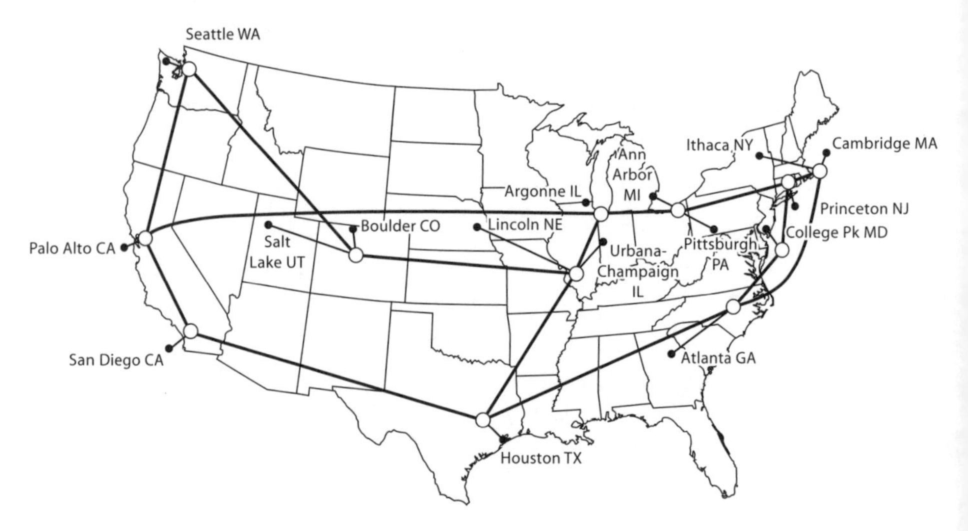

Figure 2. NSFNET high-speed Internet backbone (c. 1992) put NSF at the center of the "information superhighway." Courtesy Merit Network by CC BY-SA 3.0.

Figure 3. The Stampede supercomputer at the University of Texas at Austin is part of NSF's Extreme Science and Engineering Discovery Environment. Courtesy NSF.

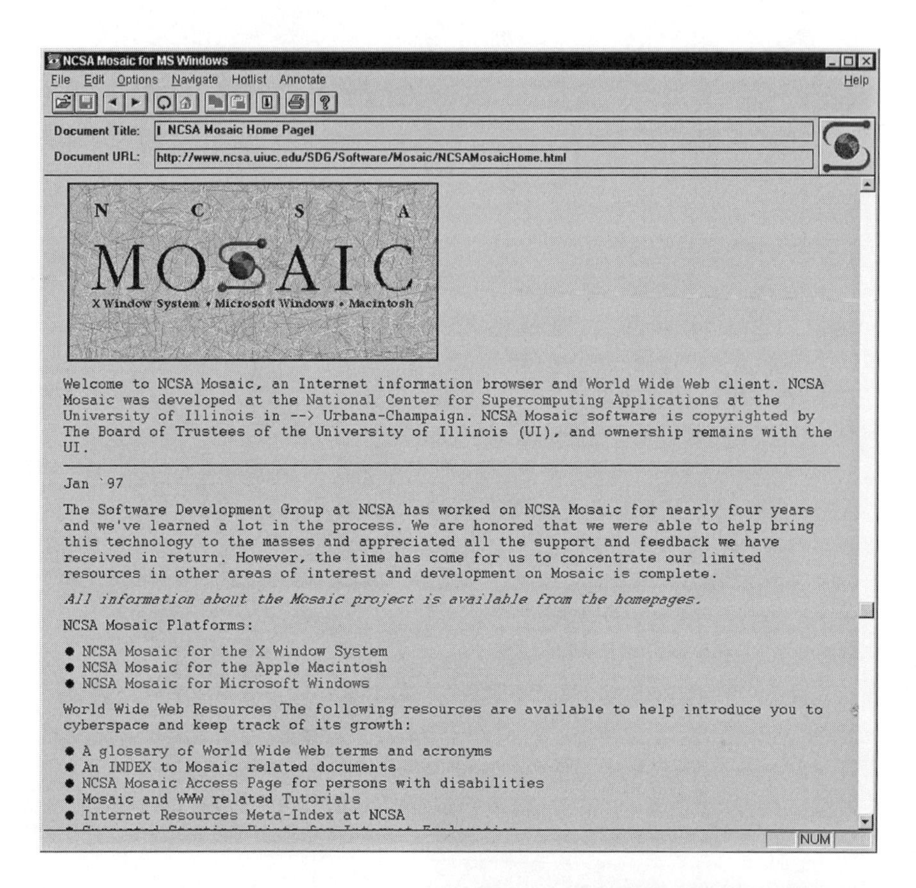

Figure 4. When he visited Illinois in the fall of 1993, NSF's Fred Wendling saw something very close to this NCSA Mosaic 1.0 version. The browser is reading a *later* webpage from 1997. From www.ncsa.illinois.edu/news/press#mosaic.

Figure 5. NSF mailroom (1996). During a federal government shutdown in January 1996, pictured here, some 40,000 pieces of mail, including thousands of proposals, piled up.

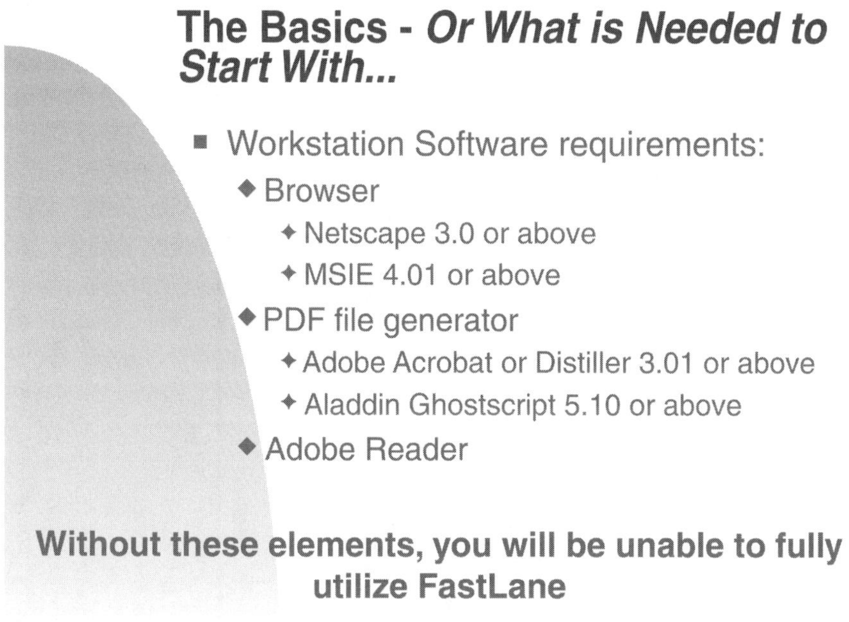

The Basics - *Or What is Needed to Start With...*

- Workstation Software requirements:
  - Browser
    - Netscape 3.0 or above
    - MSIE 4.01 or above
  - PDF file generator
    - Adobe Acrobat or Distiller 3.01 or above
    - Aladdin Ghostscript 5.10 or above
  - Adobe Reader

Without these elements, you will be unable to fully utilize FastLane

Figure 6. From UCLA's "FastLane Proposal Preparation 101" (June 2000). The university prepared 62 slides for faculty and staff, including this stern warning about FastLane's software requirements. From www.research.ucla.edu/ocga/sr2/fastln.htm.

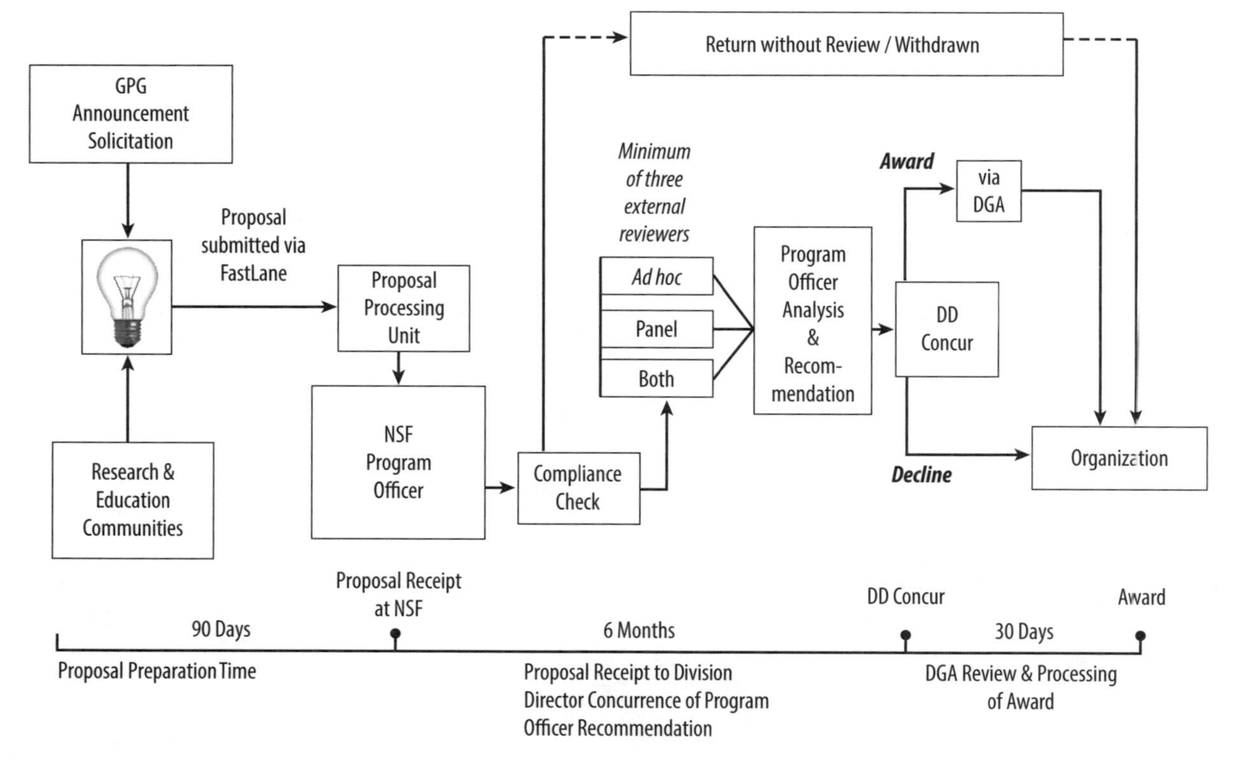

Figure 7. NSF proposal process. Ideas and research proposals originate at the left and are submitted to NSF program officers via FastLane. After passing through compliance check, review, evaluation, and "DD concur" (approval of the division director), they are either declined or funded by DGA at far right. Along the bottom of the flowchart is an ideal timeline. Redrawn from NSF, "Grant Proposal Guide" (2009), www.nsf.gov/pubs/policydocs/pappguide/nsf09_1/gpg_3ex1.pdf.

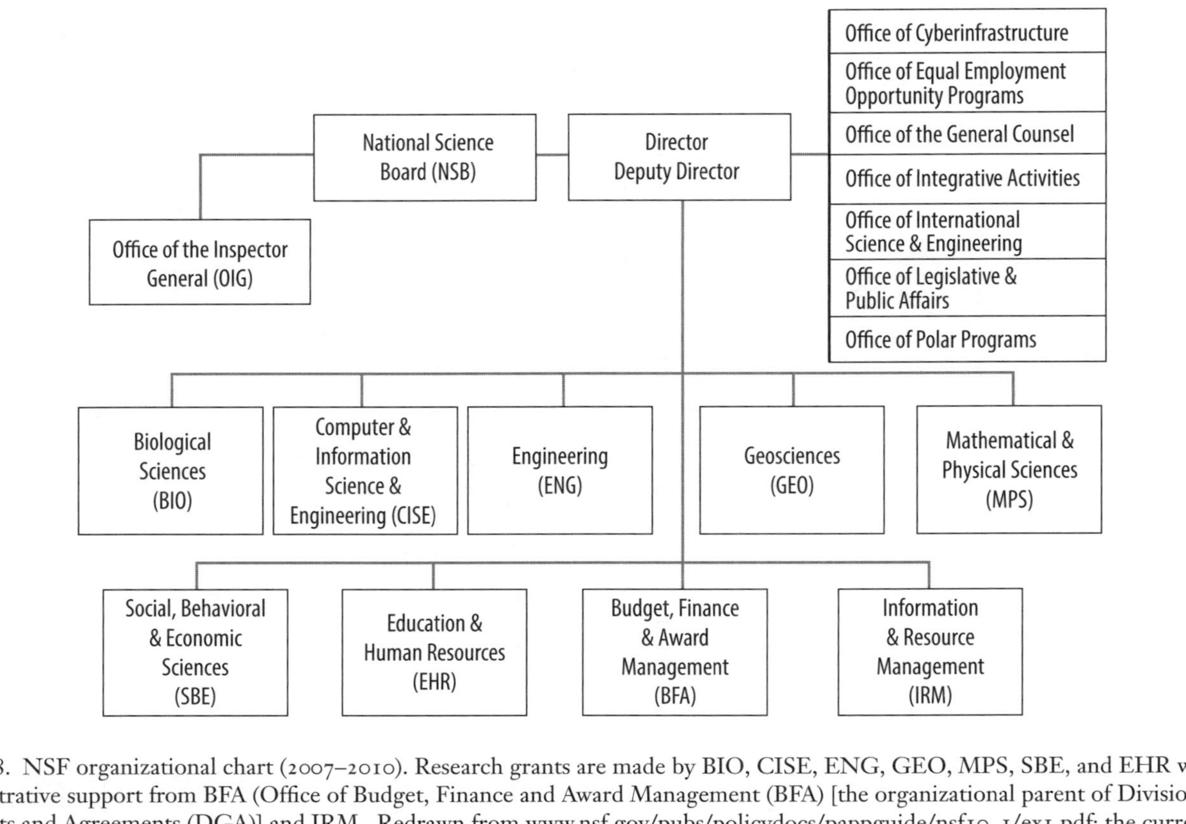

Figure 8. NSF organizational chart (2007–2010). Research grants are made by BIO, CISE, ENG, GEO, MPS, SBE, and EHR with administrative support from BFA (Office of Budget, Finance and Award Management (BFA) [the organizational parent of Division of Grants and Agreements (DGA)] and IRM. Redrawn from www.nsf.gov/pubs/policydocs/pappguide/nsf10_1/ex1.pdf; the current version is at www.nsf.gov/staff/organizational_chart.pdf.

5

Research Administrators as Lead Users

While grant coordinators and assistants at the department, laboratory, and research center levels (discussed in chapter 4) form one critical piece of the research administration system at larger universities, virtually all universities and colleges with externally funded research have a sponsored research office. This central office commonly has different names at different schools, as well as modest variations in the scale and scope of its administrative activities and services.[1] Nevertheless, there is great overlap in the core function of SROs. For proposals directed to the National Science Foundation (and most other funding agencies), the SRO is the final university authority in officially approving a proposed project's compliance (both for the school's policies and to assure adherence to the rules of the sponsored funding agency)—this is "pre-award" administration. Also, the SRO often plays a fundamental role in oversight and financial accounting of funded projects—"post-award" administration. In general, an SRO director oversees a staff of research administrators (typically consisting of grant administrators on the pre-award side and accountants on the post-award side) and reports to a top-level administrator—often the university vice-president for research or the vice-provost for research. At smaller schools, the SRO may only consist of a director and one or two administrators. SROs at major research universities may have 20 or more staff members.

SRO directors—in coordination with top university administrators—are responsible for structuring the office and supervising the grant administrators. The grant administrators often are the primary administrative interface between principal investigators and the funding agencies, and in nearly all cases they are responsible on behalf of the university for officially submitting grant proposals developed by academic researchers.[2] Some SRO staff do both pre-award and post-award administration, although at most of the schools we visited they specialize in one or the other. These administrators develop expertise with funding agencies' requirements to review grant proposals and assure compliance of research projects. At some schools, in the absence of department-level support staff, they also help PIs develop sponsored project budgets, but more often they just check to assure proper compliance with budgets (requesting revision by PIs if necessary), as well as with Institutional Review Boards, facilities requirements, and proposal length and format requirements.

SROs are critical to all universities and colleges engaged in sponsored research, though they often lack proper visibility. Faculty members or university administrators who do not actively seek sponsored research funds, assist with sponsored research at the department or center level, or oversee the SRO at top-level central administration often have little or no knowledge about the work of SROs; sometimes they are unaware that such an office exists. For research-active faculty members, these offices are essential to obtaining funding, although sometimes still underappreciated. Since many faculty interact with SRO staff during tense proposal submissions, the working conditions for most SRO staff, even at the best organized offices at the best funded universities, are unusually stressful.

We conducted interviews at 29 universities with more than six dozen SRO directors, administrators, and staff members responsible for pre-award or post-award administration of NSF grants. We also spoke with top administrators, including university vice-provosts, vice-presidents, and others overseeing SROs and developing overall research policy. We found SRO offices at various locations across university campuses: many times we walked to the margins of campus or even to a nearby but off-campus location; once in a while we found SRO staff centrally located in a specialized research complex or central administrative building. It so happened that we interviewed the two people who have the best (rival) claims for being the "first" to submit a proposal on FastLane: one of our early interviews, at Berkeley, and one of our later interviews, with a former UCLA research administrator, were both certain that theirs was the first FastLane proposal.[3]

The background and prior work experience of SRO staff varies considerably. One common observation of our interviewees is that they did *not* plan for a career in research administration; few knew the field existed until they entered it.[4]

Virtually all grant administrators are college graduates, and some (often those who have risen to associate director or director) have master's degrees in public administration or business administration; a small fraction have science PhDs. Many grant administrators brought prior experience with administrative support, oversight, or accounting for nonprofits, the military, federal or local government, corporations, or universities. A handful worked as students in SROs and moved to full-time work after graduating.[5] Successful grant administrators have unusual patience, substantial poise under pressure, and above all attention to detail that is necessary for evaluating complex compliance issues. The deadlines for grants, and faculty and academic staffs' propensity to work on refining proposals right up until the deadline, put extraordinary time and work pressures on SRO grant administrators, particularly around popular funding agency deadlines.

This chapter focuses on SRO managers and staff as lead users of NSF FastLane—how they experienced the system for pre-award and post-award administration. We assess possible differences with experiences, opportunities, or challenges at particular categories of universities (EPSCoR and HBCUs) and how FastLane and other electronic systems influenced the strategy and organization of SROs. We also discuss several new custom and commercial front-end systems that assist with electronic research administration (so-called system-to-system approaches). We visited many sponsored research offices that ran impressively well, like finely tuned machines, lending faculty at those institutions immense advantages in the highly competitive rounds of research funding; in chapter 7, we assemble some observations as "best practices" aimed at university administrators.

From Paper to FastLane

In the paper-based days—prior to FastLane's introduction in the late 1990s—SRO staff often first saw an NSF or other agency grant proposal when a PI walked it over for review. Sometimes, when a PI sought a preliminary consultation prior to an agency deadline, SRO administrators helped him or her develop a project budget; other times they just reviewed one created by the PI or a department-level assistant for correctness and compliance. While SROs have always played a fundamental role in officially submitting sponsored research projects—indeed, having the sole institutional authority to do so—the level of assistance they provided to PIs varied greatly. This was particularly true in the paper-based days.

Some SROs did all the photocopying, packaging, and express mailing of proposals, while at other institutions these tasks were the responsibility of PIs or academic department staff, or at least it was strongly encouraged that PIs or academic departments perform these necessities. Size of school or research profile had little correlation to schools that *sometimes* provided photocopying, packaging, or mailing services. However, schools that provided such services as *standard practice* tended to be smaller universities with far fewer sponsored project proposals (several hundred or fewer to all agencies per year). Among the schools we visited, University of Massachusetts Amherst was an exception. It has long had a substantial research profile (far more than several hundred proposals a year), yet the UMass SRO did all the copying, packaging, and mailing—a responsibility of clerical staff rather than the grant administrators.[6] SROs at many of the largest research universities quickly would have been overwhelmed with providing such services since they handled hundreds of proposals each week.

Despite far fewer proposals, the SROs at smaller institutions (and the added services they often provided) faced memorably hectic operations on deadline days. As the SRO director at Santa Clara University recalled, "When you had a faculty member arrive at 3:00 p.m. with one copy for final signatures, we could almost guarantee that the copy machine would break and we'd be scurrying across campus making copies. We have been known . . . to actually move furniture out because the copier would work but the sorter wouldn't, so we'd be literally sitting . . . spread out on the floor, with every available person in the office . . . frantically collating copies so we could get them stapled, into a FedEx envelope, and over to FedEx prior to the deadline." Even the final step could be a challenge, either the very circuitous route driving to the nearby Santa Clara FedEx office to meet the deadline for overnight delivery or the quicker path of walking packages straight over the railroad tracks. "The incoming trains" made the railroad route "a frowned-upon adventure."[7]

Typical of the large elite research universities, University of California, Berkeley's SRO helped PIs with copying, packaging, and mailing to a degree, but over time, with a rising tide of proposals, this level of service became "unsustainable." This SRO had to strongly encourage PIs or departments to do all the copying, collating, and stapling. That Berkeley's SRO, and most SROs nationwide, relax policies at times, to provide such services in a bind, indicates the great importance of sponsored research to universities—to further the institution's research profile and for the university to recover overhead (indirect cost funds). As such, the Berkeley SRO director characterized the paper-based days as "horrible." People would bring their 15 proposal copies for review "with hand trucks," and the SRO would have to temporarily store these boxes of proposals before reviewing and mailing them. Inevitably proposals would require last-minute changes, and PIs would "have to come over, unstaple everything, replace pages, restaple stuff—so [the SRO] had to have workspace available with staple removers and large-scale staplers." The office was "always in tight spaces . . . hallways with credenzas and supplies on top of them and supply cabinets next to them, and people hustling around each other and bumping into each other coming in and out of doors. It was not serene or tranquil."[8] At Arizona State University (ASU), the SRO would offer the "carrot" of making the photocopies if PIs got the completed proposals to them "a few days before the deadline."[9] This effectively motivated a fraction of the PIs and lessened the typical deadline crunch. ASU's SRO still had its share of hectic agency deadline days, but this policy made things a bit more manageable.

Not surprisingly, at Berkeley, ASU, and most other universities, the SROs were excited to find out about the plans for FastLane. Most SRO managers and staff learned about FastLane in the mid-to-late 1990s at National Council of University Research Administrators or Society of Research Administrators meetings. SRO directors and associate directors tended to attend these national and regional meetings regularly, as did many grant administrators on a rotating basis with their colleagues. SRO directors and staff generally had a favorable response to learning of FastLane and that it would become the mandatory system to propose research projects to NSF. Along with this enthusiasm, however, there often was some apprehension with the idea of a new (electronic) system and concerns for how it would work.

Concerns about the pending transition in the SRO were eased by half-day NSF-led FastLane workshops at national and regional NCURA and SRA meetings. These sessions also provided very useful feedback to NSF. On a pilot basis, a couple dozen SRO administrators and a smaller number of PIs had been involved in a FastLane pilot external review committee, and a handful, including those at the Berkeley, even earlier in FastLane's truly formative stages, as part of the Federal Demonstration Partnership.

The FDP, initially the Florida Demonstration Partnership, began in 1986 as an experimental collaboration of five federal agencies—NSF, National Institutes of Health (NIH), Office of Naval Research, Department of Energy (DOE), and the U.S. Department of Agriculture (USDA)—schools in the State University System of Florida, and the University of Miami to test and evaluate grant mechanisms using a standardized set of terms and conditions. In 1988 it became the Federal Demonstration Partnership and over time it grew to include 10 federal granting agencies and roughly 100 universities throughout the nation. Its mission is to advance best practices for research administration through maximizing productive resources and minimizing administrative costs. Beginning in the second half of the 1990s electronic research administration—including electronic grant submission and management by NSF's pioneering FastLane and subsequent systems from other agencies or interagency collaborations—became a focal point of FDP.[10]

As the Berkeley SRO director put it, "We were one of the original 16 FDP institutions, and so FastLane was kind of initially rolled out via FDP. So our participation, and I think UCLA's as well, both originated at the FDP." He emphasized how NSF really "encouraged feedback" at NCURA and SRA national and regional meetings. "They did a really nice job as marketing this as not an additional layer of bureaucracy, but as a tool. There was always a way to get back to them. It wasn't that you had to go the FDP meeting and get somebody in Washington to listen; you could communicate directly with the NSF [at NCURA and SRA], and they were responsive."[11] Many others SRO managers and staff, including the University of Nebraska–Lincoln's SRO assistant director at the time, echoed this sentiment concerning both FastLane training sessions at professional conferences and the FastLane Help Desk: "They'd be very interested in feedback . . . I felt they were service-oriented. They were trying to do the best they could to help with the issues."[12]

However, not all SROs had the computing technology or the inclination to be early adopters like Berkeley and UCLA. At Indiana State University, the SRO was not wired for broadband (in the recollection of one interviewee) when FastLane was first introduced in 1998. So Indiana State's SRO continued to strongly favor NSF paper proposals to avoid having to access FastLane through telephone dial-up Internet service. After leaving for a similar SRO position at the University of North Carolina at Greensboro (UNCG) shortly after FastLane became mandatory, this staff member was excited about using FastLane for the first time since UNCG had broadband Internet throughout the campus. All the same, she also had a bit of trepidation owing to limited staff resources at UNCG. "We were concerned about the extra man-hours that it would take for us to work with the system. Would it take more of our time to submit proposals using that electronic system than it would with just paper? We've always been a small office . . . There were three professionals then; there are three professionals now. That was the chief concern, but we submitted our first FastLane proposal here at Greensboro without any problems."[13] Concerns about FastLane being a time sink were common among already heavily burdened SRO staff at the launch of the system.

Experiencing FastLane

Many SRO staff who work with NSF grants—especially those at smaller schools—have responsibilities for many different federal agencies, as well as for private foundations. As such, FastLane did not represent the opportunity for SRO staff once and for all to be relieved from paper proposals, only paper proposals to NSF. Helping their universities' faculty and academic staff with FastLane passwords and learning the system themselves took some extra time, but most SRO staff members believed that these investments in time paid off strongly in greater efficiency down the road. "Greater efficiency in proposal development, simultaneous editing, and achievement of succinct and accurate forms," was one SRO officer's expectation.[14]

The vast majority of SRO staff had positive experiences when they first started using FastLane, though some identified a number of early glitches. The most common complaint was the slowness of the system—leading some SRO staff (and PIs) to label the system "SlowLane."[15] System slowdowns could be a particular challenge on the final day for a program solicitation if a grant administrator had to submit several proposals. A sponsored research administrator at University of Alaska Fairbanks Geophysical Institute (which, unusually, was a separate legal entity with its own SRO infrastructure at the time FastLane was launched) recalled they had problems in the late 1990s and early 2000s with the Arctic Research Opportunities deadline. The small staff would need to submit 20 or more individual proposals on deadline day, and he remembered the system being "very slow."[16] It was the largest volume deadlines, such as the CAREER and other cross-cutting programs, or the very largest of the specific research programs, that led to the greatest challenges—often resulting in major system crashes and the complete inability to submit (discussed more fully in chapter 4). In such situations, SRO staff often served as an intermediary between PIs and

NSF staff (program officers, FastLane Help Desk personnel, or other Division of Information Systems staff) to arrange an extension if a deadline was missed because FastLane was overloaded and temporarily inoperable.

Another challenge, also discussed at length in chapter 4, was access to Adobe Acrobat or other software to convert documents to PDF. PIs sometimes purchased this software, or their departments did, but in many cases SROs helped convert proposal documents (see figure 6). While SROs typically were the last resort to do document conversion, it was not unusual for them to end up with this time-consuming chore—particularly at smaller schools where PIs had no departmental grant preparation support staff. Problems with rendering graphs, equations, or other graphics typically were addressed by the PIs themselves (or with help from department administrators or faculty colleagues), but occasionally SRO staff would assist with this as well.

At first usage, most SRO staff found the screen design and functionality of FastLane to be basic and straightforward. "User friendly" was a common descriptor among SRO interviewees.[17] Nonetheless, some SRO staff found the system still took a while to learn and at times could be confusing—especially in figuring out how to use it for less common project or proposal structures. A grant administrator from the University of Washington recalled that when he "first did a collaborative when we weren't the lead, it confused me when it gave a warning like there's no project description, no project summary, or, I think, references because collaborations don't even have those buttons."[18] In this case, and most others, figuring out the proper way to use the system on one's own, with help from nearby colleagues or the FastLane Help Desk, made it more manageable the next time such a circumstance arose.

For SRO staff, the component of NSF proposals (and those to other agencies), of greatest focus has always been the project budget. Other responsible authorities (department chairs, deans, even vice-presidents) are signatories on NSF grant proposals—verifying the activities and research plan are appropriate for the institution. Often these individuals have too heavy a workload and too many proposals to give detailed, line-by-line attention to the budget—a responsibility generally seen as resting with the SRO. At schools with department grant administrators preparing budgets, these staff have the experience and expertise to get budgets correct and compliant most of the time. Yet SRO staff still have the responsibility to assure budgets are done properly. Using incorrect salary information, improper fringe rates, identifying the wrong number of project months for project staff, failing to adjust for annual salary increases, applying indirect cost incorrectly, and other mistakes potentially can be costly to the university, and an added hurdle for fully completing a funded project.

The shift to FastLane normally facilitated SRO staff getting access to draft budgets significantly earlier in the submission process. It also provided a platform where they could work on or do adjustments to the budget, all while eliminating or lessening concerns and problems with multiple draft budgets being worked on by the SRO and the PIs. The ability to access the budget remotely—and not to have to worry about sending and trying to share potentially incompatible spreadsheets or other software—was of great benefit to SROs, PIs, and overall to sponsored research proposal preparation.

Many SRO staff members emphasized the great value of receiving pieces of the proposal earlier with FastLane, and particularly the positive impact of receiving the budget earlier. Most believed the FastLane budget module worked well from the start. A small fraction mentioned minor issues or design flaws, the majority of which were corrected or adjusted relatively quickly to add to functionality or to make the module more user-friendly. For instance, in the first years of FastLane, the need to hit the "calculate" button to update changes in the budget, rather than just being able to use the enter key, was counterintuitive to some users—and at odds with how a growing number of web platforms automatically calculated data.[19] This feature was soon adjusted by the FastLane team to allow use of the enter key to update budget calculations. Overall, SRO staff members have been very positive about the budget module and minor adjustments to improve it over the years. The other, post-award side to budgeting and accounting using FastLane occurred when a project was funded.

The original modules of FastLane were geared to pre-award grant submission functions. Shortly thereafter, NSF rolled out post-award functionality for FastLane. One aspect of FastLane post-award—the only one PIs typically interact with—is annual and final reporting on funded NSF projects (discussed in chapter 4). The other fundamental post-award feature is the accounting and funds-transfer functionality used primarily by SRO accounting staff. Some SROs assign staff to projects through the entire life-cycle, from proposal to completion; more often, they structure the work of the office to divide staff between pre-award and post-award specialists. Post-award specialists often had accounting backgrounds, and they were using the system to incrementally draw money transfers for expenses on funded projects.[20] Virtually all of the post-award specialists spoke highly of the FastLane system for drawing down funds for projects. A University of Montana SRO post-award specialist provided a common perspective: "We draw down on a reimbursement basis, so I review our cash needs for National Science Foundation grants and then just go in and do a drawdown. It's very user friendly."[21]

Shortly after the arrival of FastLane's functionality for drawing funds electronically, universities and other awardees had the option of using a centralized federal system for wire-transfer draws on federal grants and contracts developed jointly by the Financial Management Service of the Department of Treasury and the Federal Reserve Bank of Richmond and operated by the latter. This system, Automated Standard Application for Payments (ASAP), is used widely within the federal government, including by the USDA, DOE, Environmental Protection Agency, and Department of Commerce.[22] For larger universities with sponsored research projects from many different federal departments and agencies, it offered a one-stop system. University of Washington used ASAP for NSF project drawdowns for roughly a half decade but switched back to FastLane in 2006 for drawing funds for grants. In the words of one Washington SRO post-award specialist, "FastLane was a little bit easier system; there were less screens to go into."[23] She added that it commonly took two days to receive money from NSF through FastLane and just one from ASAP.[24] This, however, stood as the one small complaint, as SRO post-award staff uniformly praised FastLane's post-award financial functionality—even more than the positive assessments by SRO pre-award staff on pre-award functionality.

HBCUs and EPSCoR schools

SROs at historically black colleges and universities generally did not experience FastLane in fundamentally different ways than other schools; nor did SROs of universities in EPSCoR states and territories. Certain programs, institutional-wide projects, and proactive outreach efforts of NSF led to minor variations in the transition to FastLane, but for the most part these schools faced similar challenges and opportunities as institutions with comparable levels of resources outside of these two categories. The five HBCUs we visited (Howard University, North Carolina A&T State University, Jackson State University, University of the Virgin Islands, and Florida A&M University) were among the most funded institutions in the HBCU category (in terms of number of NSF projects and overall NSF funds). Among the HBCUs, Howard stands out as a research institution receiving substantial NSF, NIH, and other sponsored research funding.[25] The other four schools we visited receive moderate levels of support, and a greater than typical percentage of their NSF funding comes from educational rather than research programs. Howard University and Florida A&M have moderate sized SROs (between 10 and 20 staff members), while NCA&T, Jackson State, and UVI have small SROs (six or fewer). In general, schools with smaller SROs faced greater challenges with the introduction of FastLane and, later, with other electronic grant submission and management systems.

None of the more than six dozen SRO staff we interviewed (at 29 different schools) indicated that their SRO staff size expanded as a result of the introduction of FastLane. At some SROs an existing employee was designated as or became a FastLane "guru" or specialist, but we found no evidence any were hired as such before or after the launch of the system. The presence of FastLane, and other electronic grant submission and management systems later on, unquestionably influenced the skill sets sought for future SRO hires (to be more IT oriented), but often this was little different from other administrative support positions at universities as computer and web applications proliferated across campus. As such, the IT knowledge and skills of existing staff at the time of FastLane's launch—particularly at small SROs—could make a truly critical difference to how well the office, and hence the school, adapted to the system.

A prime example was the relatively smooth adoption of FastLane at Jackson State University, owing greatly to the efforts of David Wilson who worked in the sponsored research office there. Wilson, as Jackson State's SRO director Rita Presley put it, "became the FastLane guru."[26] NSF outreach efforts for FastLane beginning in the late 1990s extended far beyond national and regional meetings of NCURA and SRA. NSF's Evelyn Baisey-Thomas, Beverly Sherman, and other DIS staff visited schools throughout the nation.[27] NSF outreach staff always strongly encouraged that sessions be open to SRO and researchers from neighboring schools.[28] As an HBCU in an EPSCoR state, Jackson State was early on NSF's radar for schools to visit for outreach and training. It has also long been part of the Mississippi Research Consortium (along with University of Mississippi, Mississippi State University, and University of Southern Mississippi), or MRC, as well as of a professional association of SROs at HBCUs, the National Sponsored Programs Administrators Alliance of Historically Black Colleges and Universities (NSPAA).[29] NSF outreach staff visited campus multiple times to conduct training, offer assistance, and get feedback on FastLane. With Jackson State's central location in the state, roughly half the MRC's FastLane meetings were held there.[30] Jackson State SRO staff also attended FastLane events at other MRC schools, and at NSPAA, NCURA, and SRA meetings. David Wilson attended several of these outreach events and, with his background, readily took to FastLane. He first attended a four-hour FastLane workshop at Mississippi State where each attendee had a PC and could practice using the system, ask questions, and offer feedback. Other similar workshops at Jackson State and elsewhere in Mississippi soon followed. Wilson, and more broadly the SRO, worked closely with NSF to provide advice on the system and to help educate the other HBCUs (more than 30 schools) in the NSPAA. Presley recalls that NSF's training specialist "Evelyn [Baisey-Thomas] was referring everybody to David Wilson at Jackson State, 'he gets it.' "[31]

Wilson had been hired in 1995 as a research data-management specialist at the SRO and for "keeping up with reporting." He had a background in PC computer sales and computer engineering. Within a year of arriving, he learned of the development of FastLane and was "intrigued with the idea of electronic submission." Recognizing his aptitude and computer skills, the SRO director assigned him to implement FastLane at Jackson State. He recalled, "We volunteered to work with NSF in the early days . . . Evelyn Baisey-Thomas was a person that we worked very closely with, from NSF FastLane . . . it got to the point where we were able to serve as . . . [one of the] model institutions that were successfully using the electronic proposal process . . . I guess one of the things that helped me . . . I did have a background with computers, and I was familiar with the Internet. It wasn't a learning curve for me."[32]

For Wilson, FastLane was easy and intuitive right from the beginning. "I do recall that even at the beginning, the first time I used it, it was something that made sense, it was put together very logically . . . it flowed very smoothly." He also thought NSF did an excellent job with adjusting design elements and functionality in the early days to make it even easier and more intuitive. "They did a wonderful job of creating user-friendly modules or user friendly versions that, in my opinion, were better with every iteration."[33]

In addition to assisting his NSPAA colleagues, David Wilson also was one of the principal trainers to help Jackson State faculty use the system and would give workshops to departments or colleges. He always made it a prerequisite for attendance for faculty to have a FastLane account already set up.[34] In some cases, this was challenging because a portion of the faculty had very dated computing equipment. As Rita Presley put it, "There may have been some instances where faculty whose PCs were outdated [and] didn't have the necessary storage [or] memory to run the system." Some faculty at Jackson State were even reluctant to use computers in FastLane's early years, or they wanted no part of FastLane. The university's SRO has long been service-oriented and would take all the files from disk (or, more recently, off a flash drive or as email attachments), convert them to PDFs, and do all the uploading if faculty requested.[35] Most faculty, however, soon embraced the system; frequent training and commitment to service by the SRO helped bring about this positive outcome.

The case of Jackson State demonstrates how one person can clearly make a difference for how the SRO and the PI community experience and adjust to FastLane. This was particularly true at smaller schools where the number of NSF grant proposals was not overwhelming—though Jackson State's sponsored research profile has grown significantly over the past decade.[36]

A far higher portion of SRO staff from HBCUs (compared with other schools) did not give us permission to make their interviews public or allow us to identify the interviewee or the school (we can only use the interview data in aggregate).[37] In some cases, this appeared to be the decision of an individual SRO staff member while in other cases it was clear that the SRO director or a higher-level research administrator stipulated restricted access for all interviews. Such caution might come about because the HBCUs have a special relationship with NSF. NSF has clearly tried to be proactive in supporting HBCUs and, above all, not disadvantaging them with the introduction of new technology or new policies.

While the other HBCUs we visited did not generally have one standout FastLane "guru" like David Wilson, most had commonalities with Jackson State. NCA&T and Florida A&M belong to NSPAA, which cooperates on FastLane and other electronic research administration issues. All HBCUs we visited had SRO staff who became proficient in using the system and helping others. A small fraction of faculty at these schools (somewhat more than at other comparably research-intensive non-HBCUs)—besides, of course, the well-resourced Howard University—lacked computing and broadband networking in their office to use FastLane when it was introduced, sometimes from lack of department resources, other times from their choice not to actively embrace computer, software, or networking technology. There were no computing or networking infrastructure deficiencies in the SROs, which was not true of all smaller non-HBCUs we learned about.[38] All the HBCU SROs we visited seemed committed to training and helping PIs with FastLane, though none appeared quite as service-oriented as Jackson State.

The implementation and early experiences of FastLane at EPSCoR schools somewhat mirrored that at HBCUs.[39] As with David Wilson, one skilled early adopter who embraced the system could make all the difference. Even while submission via FastLane was optional, as the University of Nebraska–Lincoln SRO assistant director at the time, Norm Braaten, recalled, his institution's EPSCoR proposals were required by NSF to be submitted via FastLane.[40] This trial by fire introduced Braaten to the system at an early date, so he was receptive to the request by two Nebraska faculty members to submit by FastLane to NSF programs that were not yet requiring FastLane.

As a result of Braaten's early exposure, the first Nebraska submissions using FastLane went smoothly. In the late 1990s Braaten was encouraging all faculty to use FastLane, and the 1 October 2000 requirement for submission by FastLane posed no particular challenges. "So when all proposals needed to go through FastLane, it didn't matter at all. We were probably ready by 1998, 1999 to do that. We didn't require FastLane use at that time, but we could have, had we needed to. By 2000, it was nothing, it was a piece of cake."[41]

Mark Ruffolo had a different perspective on FastLane's introduction at South Dakota State University (SDSU), where he came to lead the SRO in 1999 and now is associate vice president for research.[42] Ruffolo believed most faculty would want to continue using paper proposal submissions during the optional period and did not push them to do differently. Ruffolo thought the difference between SDSU and Nebraska may have been "cultural," or the way Braaten "promoted" FastLane.[43] As a PhD scientist, and PI on multiple research grants before entering research administration, Ruffolo believed he saw things more from the PIs' vantage point, where what might be presented by the SRO as services could conceivably seem to a PI like relinquishing control—something many independent-minded faculty resist. As Ruffolo stated with regard to the paper-based pre-FastLane days, "Personally, I've never tried to interfere as a research administrator with the process of reproducing the number of copies, and in fact as it turns out, we always had the investigators submit the proposals. They had responsibility for the delivery of the proposals, and they've always welcomed that."[44]

In comparing these two colleagues, one sees different personalities and educational backgrounds—in Braaten an MBA with strong computer skills focused on customer service and in Ruffolo a scientist and senior administrator with an understanding of investigators' general desire for autonomy and control over their research and applications for funding. At schools with small SROs, including nearly all HBCUs (excepting Howard and Florida A&M, both of which have mid-sized SROs) and most EPSCoR state schools, the skills and preferences of existing SRO managers and staff, and the computing and networking infrastructure (both in the SRO and throughout campus) at the launch of FastLane made a meaningful difference in how, and how early, it was implemented.

Another key factor in FastLane readiness and adoption was NSF staff campus visits. While NSF targeted both HBCUs and EPSCoR schools, it was easier to go to HBCUs with a meaningful research profile (fewer than a dozen schools) than to the hundreds of EPSCoR schools. At the EPSCoR schools we visited, interviewees regularly mentioned that EPSCoR block grants helped to upgrade computing capacity and build up research offices and research infrastructure.[45] And, across the board, most larger universities with large SROs fostered and encouraged use of FastLane during the optional period, so the mandatory submission via FastLane requirement rarely posed problems—other than the FastLane system overloads and PDF challenges, which lessened over time.

Strategy, organization, and internal ERA systems

How SROs organize their offices and the strategies they adopt can affect their entire institutions' interactions and experiences with FastLane, as well as with other electronic grant submission and management systems. The strategy, organization, and policies of SROs often change with a new director or upper-level research administrators, and sometimes they evolve in response to focusing events and new technologies. Over the past two decades, there have been many changes in the research administration environment. On the heels of FastLane, other federal departments and agencies have implemented their own electronic grant submission and management systems—from NASA's NSPIRES to the Department of Defense's CDMRP eReceipt Online Proposal Submission System. In 1999, Congress passed the Federal Financial Assistance Management Improvement Act to simplify management of federal grant programs. In 2002, Grants.gov was launched in an effort to make learning about and applying for federal grants simpler and more uniform. Grants.gov from the start has been a multi-agency portal (listing NSF grant opportunities and those of many other federal agencies). Later in the decade it became controversial as federal mandates called on it to become the standard for all federal grant submission. Grants.gov became an option for submitting proposals to NSF programs, even as it became mandatory for grant submission to departments and agencies such as NIH and USDA. These developments have deeply affected university SROs and some

PIs. Another event of the past half dozen years affecting SROs (and universities more broadly) was the financial crisis of 2007–2008 and its continuing effects.

The financial crisis reverberated throughout higher education, but it disproportionately hurt state schools, which to varying degrees depended on state funds. Among public universities, institutions in states with the greatest financial troubles generally faced the deepest challenges. Arizona was one such state due its massive declines in real estate values, second only to Nevada. Several years into the financial crisis, Arizona's budgetary shortfall was 39 percent, the third worst in the nation.[46] The two major research institutions in Arizona in recent years have adopted different strategies with regard to SRO organization and PI pre-award support. The University of Arizona has retrenched in certain areas, which has included a declining number of positions for pre-award support locally for PIs within some departments. Centrally, the university's SRO has grown—on the post-award side to meet added responsibilities—but the pre-award staff has remained relatively small (a manager and four staff members) for a school with substantial federally sponsored research funding.[47]

By contrast, Arizona State University's SRO has invested in sponsored research infrastructure since the mid-2000s, developing a decentralized structure to place staff locally in departments and colleges to aid with proposal preparation. An Arizona State site supervisor of sponsored project services explained, "When I first started working in the central office there were five people who did pre-award . . . And then we moved toward a model where we were assigned specific departments and a couple of other people were added . . . And then, a few years later, they set up . . . site offices and satellite offices . . . because it was felt that if we worked closer physically to our departments that would increase communication and would foster good working relationships . . . and it really did do all of those things."[48] The SRO, to get more people resident with local units, also reconfigured staff so that officers would perform pre-award and post-award administration, working on projects "cradle to grave." Arizona State's strategy of having satellite research offices around campus and encouraging the hiring of "research advancement" staff at departments and centers has directly counteracted what many cited with regard to FastLane and other electronic systems: depersonalizing communication and reducing or eliminating face time.[49] While Arizona State has always had a smaller amount of total federal research dollars than University of Arizona, in recent years it has been growing its research funding at a far faster rate.[50]

Grants.gov was a response to the real and anticipated proliferation of electronic grant submission and management systems within the federal government. For many SRO staff we spoke with, Grants.gov had caused rather than solved problems. In their view, Grants.gov took far more time to use than FastLane, and there was greater uncertainty with the system. Because it took more time to use, there often was less time flexibility in receiving proposals from PIs.

While most SROs have long had standard deadlines for PIs to get completed proposals to their offices prior to the agency deadline—two, three, or five days being the most common—asking SRO staff about deadlines often resulted in a chuckle or roll of the eyes. These deadlines were often treated by PIs as mere suggestions. At most, SROs would not guarantee a proposal would be submitted if the faculty member sent it in after the SRO deadline; they rarely, if ever, would refuse to accept one if there was any possibility of making an agency deadline and typically made every effort (sometimes heroic ones) to get late-arriving proposals submitted on time.[51] The general consensus was that FastLane had no impact on official deadlines and even may have reduced required lead time on unofficial ones. Not needing to get proposals copied and mailed encouraged some PIs to push the limit and make their full proposal ready only hours or minutes before a NSF deadline. Some SROs indicated that with Grants.gov, longer deadlines were needed or they had to be more strictly enforced because of the time-consuming challenges in using that system. As such, many at SROs asked some variation of the question: why couldn't the federal government just use FastLane as the standard?

The question may not be as simple as it looks. FastLane was created for NSF's particular needs, priorities, and institutional values; the great challenge with any standardized system is the difference between the different federal departments and agencies in what they require and in what form. FastLane was designed specifically for the policy and practices of NSF, as we have emphasized. While the consensus opinion among SRO managers and staff was not complimentary of Grants.gov, some indicated that it had had problems and glitches in its first years but has recently operated more smoothly. Others stressed a steep and sometimes daunting learning curve. Most SRO interviewees evaluated Grants.gov on its ease of use, for themselves, for colleagues, and for PIs. Only a very small number of focused on its potential as a solution to the proliferation of research-management systems.

Staff at the University of Hawaii at Manoa's SRO in particular called attention to the problem of proliferating systems, especially for schools with smaller to mid-sized research offices such as theirs, which employs four managers and thirteen professional staff.[52] As the Hawaii SRO director stated, "We found it very cumbersome that we had to use so many different electronic submitting systems. It's not to say FastLane is good or Grants.gov is good or NSPIRES is good or whatever. I don't really worry about good or not. I'm more worried about, just give us one system to use for all the federal government . . . That's what we want."[53]

The director spotlighted the burden that needing to employ so many different systems can impose on PIs and SROs. She also indicated the importance of front- and back-end electronic research administration (ERA) software modules that can connect to federal government systems. Potentially these modules can simplify and standardize the input of data and thus lend efficiency to pre-award and post-award processes. When we visited the University of Hawaii in late 2011 it was early in the process of implementing one such standard software, Kuali Coeus. This system is comprised of open-source, consortium-developed software modules drawing on the MIT Coeus system developed in the 1990s and refined in the early 2000s.[54] Hawaii's director explained, "We're going through this conversion, to do system-to-system through Grants.gov. Not through FastLane, not through NSPIRES, not through anything else. Hence, this creates a lot of burdensome work for the central office to do. And PIs were not happy about that—because it is not system-to-system through FastLane. So that has been an issue for us. And I have been trying to talk to NSF about this." She indicated that after multiple attempts, the NSF "policy office" called and told her that while system-to-system cannot be done through FastLane per se, it can be done using Research.gov, a collaborative effort led by NSF, NASA, and others that has already replaced the reporting functionality of FastLane and is expected to fully replace FastLane over time.[55] (Research.gov is further discussed in chapter 6.)

To some PIs at University of Hawaii, the news of the standardization around the Kuali Coeus front-end module was disconcerting, if not upsetting. Kuali Coeus, branded at the University of Hawaii as myGrant, handles all internal electronic paperwork, routings, approvals, and system-to-system functionality to Grants.gov. To take advantage of myGrant's system-to-system functionality, researchers will need to submit proposals to NSF not via FastLane but instead through Grants.gov. To continue to submit via FastLane would involve the complete redundancy of entering information in myGrant (required for University of Hawaii research office administration) and in a separate FastLane proposal. In effect, with myGrant the University of Hawaii SRO is looking to migrate the institution away from FastLane to take advantage of the system-to-system capabilities of myGrant with Grants.gov—though it remains open to any system-to-system functionality if that were possible for proposals with either FastLane or Research.gov in the future.

Some University of Hawaii PIs we spoke with, especially those who only (or primarily) submitted proposals to NSF, were disappointed with the news they would have to learn and use a new system. They hoped they would not have to move away from FastLane. According to a civil engineer who had recently attended the mandatory myGrant training, myGrant "looks a little daunting, I have to say. And if I were given the choice of using that or FastLane, I would choose FastLane any day. But I'm not sure I'm going to have a choice in the future."[56] An atmospheric scientist who missed all of the prior training sessions for myGrant because he was in the field conveyed similar apprehension, stating that he'll soon attend training and "will find out what new monster has been created at the university. Is it going to function well with NSF, or not? And I keep my fingers crossed."[57]

Kuali Coeus, or myGrant, is not the first front-end system at University of Hawaii, which has used both commercial software products and unique contractor-developed systems in the past. What is different with myGrant is its integration to work with one proposal-submission system, Grants.gov, and not others, such as FastLane or NSPIRES.[58] The situation with myGrant at University of Hawaii demonstrates the asymmetries of experiences with grant-submission systems between PIs and SROs, especially where those offices are smaller to mid-sized. A minority of PIs submit regularly to many different agencies, but most researchers target one or just a few. Many scientists and engineers submitting regularly to NSF do not do health-sciences research (where NIH dominates) and have never used Grants.gov. If they use more than just FastLane, it might be NSPIRES for NASA grant proposals.[59] They have mastered FastLane and want nothing to do with Grants.gov, particularly after horror stories they've heard from colleagues over the years. The proliferation of systems is not a major issue for them. For smaller to mid-sized SROs, whether they assign grant administrators by departments/colleges or by agencies, each administrator has to work with numerous systems.[60] Larger universities have sufficient proposal volume for research administrators to specialize with NSF or DOE or NIH or Grants.gov. Even if research administrators are assigned to individual departments, each department tends to have small concentrations of funding agencies, and if there are multiple administrators for a group of departments, they can still specialize by agency or system.

System-to-system integration clearly is an important goal for SROs, and it likely will be a growing trend throughout research management. While Kuali Coeus is an open-source set of tools currently or soon to be in use at roughly a dozen schools, far more use schools use commercial systems. The most prevalent one is Cayuse, which is also developed to work with Grants.gov.[61] Because of the much higher volume of NIH than NSF, and the fact that other departments and agencies besides NIH use Grants.gov as their standard, commercial vendors will continue to focus on Grants.gov system-to-system integration first, if not exclusively. What is virtually certain is that SROs will continue to adopt front-end systems. Such systems allow them to move toward paperless administration and more efficiently handle ever higher volumes of proposals without equivalent expansions of staff (and sometimes with stagnant or even declining staff size over time).

Recommendations and lessons for cyberinfrastructures

SRO staff offered great praise for FastLane, both for its pioneering role with electronic grant submission and its stability and user-friendliness. Like PIs, often without prompting, they compared and contrasted FastLane and Grants.gov—characterizing the former as easy, simple, and quick to use and the latter as confusing, unstable, and difficult to use. Those critiquing Grants.gov varied in whether they restricted their harsh critique to the first years of Grants.gov or continued that characterization to the present—though most believed it has improved meaningfully over time. One of the most cited lessons for effective cyberinfrastructures was early and continuing involvement and feedback from users. This is something that NSF sought to do and, by most SRO staff accounts, did quite successfully. NSF's outreach to campuses was particularly effective in attracting SRO staff members and department-level grant administrators. The interactions between NSF and SRO staff at the Federal Demonstration Partnership and at NCURA and SRA national and regional meetings made a tremendous difference. Not surprisingly, the earliest glitches with FastLane—PDF conversion problems and server overloads and crashes—were elements that SRO staff could not have readily foreseen. PIs generally had the responsibility for converting documents to PDF, and no one anticipated the system overloads around deadlines until they occurred. NSF attempted to overshoot in needed capacity, but time and again in the early days these efforts proved inadequate.[62] SRO staff and PIs conveyed to NSF the scale and scope of this system capacity problem, which was resolved relatively quickly.

A minority of SRO staff laced their praise for FastLane with a particular recommendation or two—generally tools or functionality of FastLane (or the NSF awards database) to make their work as research administrators a bit easier. For instance, an SRO staff member from Stanford University stressed that since many of her PIs have current and pending support only from NSF, it would be extremely useful if FastLane captured and populated NSF "current and pending support" in a document so she did not need to do this repetitive task.[63] A research administrator at University of Montana offered that FastLane's functionality for finding previous applications and award documents could be improved, that in terms of ERA systems, "NIH's Commons setup is much better . . . because their setup . . . keeps everything together . . . [but with] FastLane, you've got to look for things piecemeal."[64] The assistant director for pre-award at Purdue University recommended that administrative tools for oversight of report completion be improved: "It could be a whole lot nicer if you could just look for a certain faculty member and get a list of all their . . . projects and what reports they have due . . . It gets a little bit confusing and difficult to find what you're looking for."[65] A member of Florida State University's SRO staff wished "that the person's [PIs] name was on the check sheet" and that there could be a "temporary ID number on the cover sheet" from the beginning (even before submission).[66] And finally, an administrator from NYU echoed and expanded on the issue of fields and data availability, stating it is "limited on what information it pulls for you. Usually, it's the ID, last name and a title, and the amount that was funded. But you don't have the start/stop dates, you don't know if there are subcontracts in it. These are the kinds of things you're looking for."[67]

Most SRO staff, all the same, commended FastLane for being straightforward and effective. That NSF achieved this was largely due to dedicated staff and contractors, coupled with a deep commitment to getting user feedback from the earliest stages and being responsive to this feedback in modifying the system's design and functionality. As the associate SRO director at the University of Washington, one of the 16 pilot universities working with NSF at the start on FastLane, summed up, "I . . . worked with the faculty in the department who put together a test proposal and then routed it to our office from the Office of Sponsored Programs and made sure that the electronic transmission worked successfully. And there are glitches at times, but FastLane really listened to the user and they made the system for the user. And . . . that's how it's been successful."[68]

As many SRO staff pointed out, this explicit attention to users is how FastLane differed from the rollout and early history of other systems, particularly Grants.gov. The executive director of Boise State University's SRO stated, "If an agency is going to roll something out, they need to understand who the users are, what their culture is, so that the appropriate interaction takes place. And to make sure technically it works. Because with Grants.gov technically it just didn't work. On huge submission days, forget it. You were not going to get your proposal in." Boise State's SRO helped limit problems by requesting proposals for Grants.gov fully 10 days early, but this did not solve everything. Despite that "people actually complied," they "still had last-minute problems."[69]

Unlike PIs, who generally work on FastLane only on a monthly or annual basis, many SRO staff work with FastLane and other grant-submission systems every day of their working career. The effectiveness of a system, or the problems with one, directly shapes their work experience and affects their lives. An Arizona State University research administrator with more than 25 years of experience captured the general sentiment of many SRO staff in stating, "When I first started using FastLane and learned how to use it, I would find I had a big smile on my face . . . it was easy to work with, it was very easy to understand, and I just really got a sense of accomplishment in working with it and working with it successfully. So . . . early on when I was new to it, I loved working with it; and as an experienced administrator, I continue to love working with it."[70]

6

NSF Staff as Legacy Users

While PIs and SRO staff were the primary external users of FastLane, from the start of the project the aspiration was to go paperless within NSF as well. What individuals outside NSF generally understand as the totality of FastLane was, in fact, only the modules for grant submission, proposal review, and post-award grants management. For the design and development of these external-facing modules, NSF was able to draw on its Research and Related as well as Educational and Human Resources funds, beyond the budget traditionally slated for the Division of Information Systems to create, acquire, and maintain information technology for the foundation.[1] In sharp contrast, the *internal-facing* infrastructure required to go paperless within NSF (the back-office side of FastLane) needed to draw exclusively on the much more heavily restricted administrative funds designated for DIS (see figure 8). Significant consequences resulted from this seemingly insignificant institutional practice.

As a federal granting agency, NSF is judged by the president, the Congress, the press, and the public on how well it gets dollars out the door to fund pathbreaking scientific research—to advance science and technology knowledge, communities of researchers, educational resources, and applications for the betterment of the nation and the world. Only 5 to 6 percent of NSF's total budget has customarily gone to administration, and only a relatively small portion of that to information technology systems. NSF takes substantial pride in the fact that 19 dollars of every 20 in its annual budget go directly to fund external research.[2] By some accounts, the small and varying amount left over after all other administration expenses are covered goes to DIS. The vast majority of NSF's administrative expenditures go to the salary and benefits of the agency's roughly 1,200 full-time employees, a number that has risen a bit more than 10 percent over the past three decades despite proposal volume increasing at a far greater pace.[3] The lion's share of these salary and benefit expenditures pay for program officers (in recent years the largest class of employees at NSF) and top-level managers at the agency.

The diffusion of computing technology throughout NSF under Erich Bloch's leadership from the mid-1980s to 1990 led to shifts in work culture and responsibilities, as well as in the mix of staff. These trends gathered speed in the years around 2000 with the agency's effort to go paperless with merit review and record keeping for all proposals. Prior to, and even well into, the FastLane era, proposals and all supporting documentation were kept in cardboard binders (with metal clasps on both inner sides) referred to as proposal "jackets." These physical jackets were the agency's principal information unit; they kept the official copy of the proposal on one side and supporting documentation (reviews, correspondence, memos, approvals, annual and final reports, press clippings, and so forth) on the other side of the fold. From the very start of the FastLane project, the planned internal FastLane infrastructure was dubbed Electronic Jacket, or more commonly eJacket—a system for electronic records for all aspects of a jacket that were previously maintained with paper.[4] With broad-based computerization at NSF by the late 1980s and the arrival of the early functionality and deployment of eJacket for declinations (proposal declines) in the 2003 to 2004 period, NSF, which once had more support staff than any other class of employee, steadily shifted through attrition to far fewer support staff and modestly more (higher-paid) program officers to address its ever-rising proposal volume.[5]

NSF program officers—with titles such as program director, assistant program director, program manager, assistant program manager, and the like—conduct the merit review process (by panel, ad hoc reviews, or both), evaluate proposals and review results, and make recommendations on awards and declinations (see figures 7 and 8). In the next step of the process, senior administrators—division directors who oversee different NSF research granting programs in various categories of disciplines—concur, modify, or deny recommendations of program officers. Recommendations for award with division director "concur" go to the granting body of NSF, the Division of Grants and Agreements (DGA), to finalize the award decision and set up the funds transfers to awardee institutions (or, less commonly, to unaffiliated individuals receiving awards).

Proposal declinations never get forwarded to DGA, and the administrative process is much shorter and simpler. As such, eJacket was instituted for declinations (more than 70 percent of proposals in recent years) long before the more challenging modules were created to create a paperless infrastructure for awards.[6]

Another component of the larger FastLane system was using computer technology to replace paper for panel reviews (reviewers coming to NSF or a nearby hotel for several days) or ad hoc (mailed-in individual) reviews. Initial goals were for the development and deployment of both parts of eJacket (for awards and declinations), an electronic panel system, and electronic ad hoc reviews to rapidly follow the launch of the external-facing portion of FastLane. In reality, only the relatively straightforward functionality for electronic ad hoc review (simple transfer of electronic documents) arrived alongside the introduction of FastLane. Given PDF rendering challenges (discussed in chapter 4), some PIs' requests for hand-inserted graphics, and some reviewers not wanting to read proposals on a computer screen, NSF continued to handle some ad hoc reviews through postal mailings. A rudimentary electronic panel system was launched fairly quickly, but it took several years to get the hardware and software stabilized and streamlined. With eJacket, the functionality to handle declinations came more than half decade after the launch of FastLane to the research community and three years after FastLane was mandatory for submissions. The toughest piece of the puzzle, using eJacket to make awards, only became operational and standard in 2010, over a decade after the external side of FastLane was in common use.[7]

This chapter explores how paper persisted within NSF, the development of and program officers' experiences with eJacket for declinations and awards, the human resources and organizational challenges with eJacket's introduction, and program officer and PI perspectives on the electronic Interactive Panel System. It also highlights the resiliency, skill, and innovation of program officers to create needed computing tools, as well as the steps toward next-generation services to replace—with the new Research.gov—the stable but dated external-facing portion of the FastLane system. It draws from 48 interviews with NSF program officers, and additional interviews spanning the personnel of the organization—from entry-level support staff to top management (including leaders in DIS and DGA as well, as the former NSF chief information officer and former acting executive officer of the National Science Board).

The persistence of paper

Tales of the paper avalanche within the University of California, Berkeley, and other SROs just prior to agency deadlines for major programs (explored in chapter 5) pale in comparison to what routinely was faced within the walls of NSF. For decades, NSF received tens of thousands of boxes annually, each containing the required 15 or more copies of a proposal. NSF's enormous mail volume justified its own zip code.[8] Popular deadlines, such as the agency-wide CAREER program for early-stage researchers, posed particular challenges. The movement of paper within NSF required large teams of support staff, from mailroom clerks to secretaries and administrative assistants who worked for and alongside program officers (see figure 5). Prior to FastLane, a PI's designation of an NSF program in a proposal did not result in immediate routing to that program; it took some time for proposals to be sorted and delivered and for program officers to begin to work through them. An administrator recalls that for many program officers "seeing the new ideas" as they were first delivered in newly arrived proposals was "one of the most fun aspects of their job." Some were even a bit overzealous in aggressively identifying, and in consequence being the one to claim, the most fascinating new research. With regard to those overseeing programs that crossed areas, "we had several program officers over the years that were notorious for being there right when the proposals were received to make sure they got the proposals from the up-and-coming scientists, even if it didn't quite fit their area."[9]

A far greater problem than program officers jockeying for promising proposals was the sheer physical burden of the paper. For at least a half decade after the launch of FastLane, NSF printed all electronic proposals. In this instance, the light-hearted quip about the "electronic office" in reality being the "printing office" was literally true. The agency continued to create paper jackets and manage review processes just as it had in the paper-based submission days. Some program officers and support staff we spoke with discussed the problem of paper jackets being temporarily mislaid and the great stress and significant time spent searching for them.[10] NSF had a good filing system, and program officers and other staff were officially forbidden from taking proposals or jackets outside the building, but given the immense volume, jackets inevitably were misplaced at times.[11] When seeking cross-directorate co-funding, program officers would circulate to their peers the unique instances of proposal jackets; these proposals and their documentation could be lost "for weeks, sometimes months" in the large piles on someone's desk.[12] NSF's increasing focus on interdisciplinary and cross-directorate research programs was yet another spur to going fully electronic with its proposal jackets, since an electronic instance of a jacket could be routed to a colleague without fear of its being mislaid.

The merit review process also presented hurdles in the paper-based days. Tracking down reviewers (for addresses and to assure it was the right person with a given name) often was a hefty chore prior to the World Wide Web—the PI and Reviewer System (PARS), was a helpful database to many program officers, though some found it "quirky," "idiosyncratic," and not always reliable.[13] Prior to NSF's move to Arlington, Virginia, some program officers would walk over to nearby George Washington University's library to consult reference indexes and identify potential reviewers and their affiliations.[14]

Panel reviews, particularly for the large programs, created further challenges in the paper-based days. At that time, panel reviews took place on-site over the course of several days. Larger programs held their panel reviews at nearby hotels, as there was insufficient space at NSF.[15] One large program, Instrumentation of Laboratory Improvement (ILI) received over 2,000 proposals per annual cycle and regularly used the Doubletree Hotel in nearby Crystal City to host reviews. With this program, 800 reviewers would come in for three and a half days. After a general meeting for all participants in the hotel's large ballroom, reviewers split into their panels (typically a half dozen people) to meet in regular hotel rooms with the bed removed and a circular table set up. Prior to the early 1990s, reviewers typically would hand write reviews using carbon paper for copies. A contractor was always hired to sort and organize all the paper, and to set up an industrial-sized shredder in the hotel basement to destroy all copies of proposals once reviews were completed.[16]

By the mid-to-late 1990s NSF-supplied computers (rented from a contractor) came into standard use for reviewers to write reviews and print them out. One year, near the turn of the millennium, a mistake was made at the ILI panel event and all the *reviews*, rather than proposals, were shredded, an error only discovered a week later. Some reviews were recovered from diskettes, but many others were permanently lost and a second review session had to be set up.[17] Needless to say, even in the vast majority of cases where catastrophe was avoided, the process was wearisome for program officers, reviewers, and contractors. Workloads were heavy, and reviewers read proposals late into the evening. Some panelists claimed to be sleeping only "two hours a night."[18]

Once reviews were collected, after panels or ad hoc reviews were mailed to

NSF, program officers or support staff had to remove the names of reviewers or other identifying or inappropriate remarks by cutting and photocopying. As one program officer recalled, "If you are dealing with a whole bunch of paper sometimes the names would slip through or there'd be a mistake made with getting the reviews anonymously to the right people."[19] Such problems, while not common, were quite serious as they threatened reviewers' faith in NSF maintaining the confidentiality of their reviews and thus the integrity of the single-blind merit review process. Clearly, an electronic panel review system could help ease some of the pain for program officers and reviewers alike, while also lessening the risk of human error.

Interactive Panel System

Once the proposal-submission and post-award modules (reporting and funds transfer) for FastLane were launched and stable, computerizing the panel review process and the back-office proposal administrative functions, or eJacket, were both priorities. Creating the first iterations of a complete electronic panel review system, however, moved far more rapidly than eJacket. Panel review was less complicated from a technical and policy standpoint. In addition, because panel review was an interface with the research community, it was externally facing and so could be prioritized for funding. The system was developed shortly after FastLane became mandatory for submission.

There are multiple reasons why NSF has shifted increasingly toward a panel review model over the years, even though ad hoc reviews continue to be used heavily by a minority of programs, and some use a mixture of the two review types. One reason for the growing use of panel reviews is that the reviewer base is drawn on more frequently and the percentage willing to regularly do ad hoc reviews—and do them attentively—has been difficult to sustain. Panels are often a better means to grab the attention of busy academics and get substantial and focused work out of the review community. A second reason—the basis for launching panels in the first place—is that the in-person interactivity and discussion can potentially produce better results. Panel summaries can often provide clearer, and more defensible, guidance for program officers to make decision recommendations. The first electronic panel system created at NSF's Arlington offices consisted of rooms with large and obtrusive personal computers and monitors built in to conference tables. Unlike the FastLane proposal system, we found no evidence that there was extensive involvement of end users (the research community, or for that matter, even nearby program officers) to test and provide feedback at the design stage for these dedicated panel rooms.[20] As one program officer recalled, "We dismantled those stationary computers because it was just too static . . . people would . . . sort of peer around their screens to see other people . . . so it could be a bit awkward."[21]

Shortly following this initial misstep, NSF shifted to using laptops (initially provided by NSF for reviewers, though over time reviewers commonly brought their own). The software and networking for NSF's reformulated Interactive Panel System worked quite well, by most accounts. Having access to the other panelists' reviews on screen proved highly beneficial. It facilitated reviewers' discussions and the creation of quality panel summaries. With this system, NSF also switched to sending proposals electronically (or by paper on request) in advance and holding shorter panels—a day, a day and a half, or two days each being common—with multiday marathons becoming largely a thing of the past. Reviewers could enter their reviews before they traveled to NSF, and NSF saved funds by lowering expense stipends. According to one program officer, "It improved the collaboration during the panel meeting because people could feed comments back and forth to one another through FastLane itself."[22]

The system was not without glitches, of course. A program officer emphasized that she had to remind people to use Microsoft Word to create text because the Interactive Panel System's text editor had the "bad habit of refreshing and then deleting everyone's stuff because no one remembers to push the 'Don't Refresh' button."[23] Dozens of PIs we interviewed commented on the NSF Interactive Panel System. Some cited early challenges, but the vast majority spoke favorably about the system and stressed that the technology contributed positively to the panel experience. They cited it as "seamless," "convenient," and "very easy to use."[24] In short, the Interactive Panel System—used by program officers and the research community of reviewers—closely paralleled other external-facing components of FastLane. Like FastLane's proposal-submission components, the Interactive Panel System was plagued with some modest early glitches and involved a bit of a learning curve for users, but it quickly became a stable, robust, and user-friendly system. By most accounts, neither the speed of development nor the strong degree of user-friendliness was achieved with the internally facing modules of eJacket.

eJacket

For many years, it was a commonplace at NSF that eJacket was "just around the corner." As the grant proposal and post-award grant management side of FastLane was created, refined, and launched in the mid-to-late 1990s, the Division of Information Systems and its contractors were also at work on developing "Electronic Jacket."[25] Electronic Jacket and its successors were, like FastLane, a series of modules developed with the intention of fully replacing paper with electronic documents and functionality. A prototype electronic jacket system of the late 1990s, however, was incomplete, plagued with problems, and never fully deployed. Shortly after FastLane's external side became mandatory for grant submission in October 2000, DIS's work heightened to produce the internal side of FastLane, a system to make NSF paperless for its grant proposal evaluation, recommendations, declinations and awards management, and related record-keeping functions. With the terms "e-commerce," "e-government," and the like gaining currency in the American lexicon in the late 1990s and early 2000s, the name of the new system under development was shortened to eJacket, or, as many around NSF would informally call it, "EJ."

As with the external side to FastLane, NSF's Policy Office, the Office of the General Counsel, and the Division of Information Systems were heavily involved from the start of eJacket to assure the system met the agency's needs, as well as wider federal regulations. DIS made efforts to get different categories of users involved, with partial success. DIS managed to get all categories of users represented to a degree, but they generally had a far easier time getting administrative support staff to sign on for advisory committee work than the group of employees that would become the system's primary users, program officers and division directors.

The Policy Office had a special level of involvement because its managers recognized that every detail of the system and its functionality would have policy implications. The leaders of this office had long had a positive working relationship with DIS and wanted to assure that the emerging electronic system *implemented* NSF policies rather than *dictated* policies. In the words of the longtime head of the Policy Office, Jean Feldman, "We had to really work with programs to make sure that the rollout was effective and efficient. We had to involve the people from our records retention . . . in the Division of Administrative Services to make sure when . . . we [could] . . . get rid of something, when do we have to keep something. We had to make sure that the electronic system handled all the documents necessary to be included in an electronic system. We had a heavy role."[26]

As with most other branches of the federal government, NSF has always carefully and extensively considered, and formally outlined, its policies. Unlike many departments and federal agencies, however, NSF draws many of its program officers from universities. While universities are also awash with policy, they tend to have fewer formal procedures than NSF. Many academicians are highly independent, and it is the case that informally evolving departmental practices can differ from university-wide policies. Not surprisingly, given their backgrounds, NSF program officers typically carry with them a sense of organization and management preferences that they see as best for their particular field—and for their particular research program.

NSF programs range from those that provide very large ongoing grants to create and sustain major research centers (like the National Center for Supercomputing Applications at the University of Illinois) and those focused on small businesses to programs to advance undergraduate science and engineering education or to bring basic computer networking and other funding to tribal colleges and universities.[27] Even programs focused on the common goal of advancing scientific research differ markedly depending on the scientific discipline and the varying imperatives of fieldwork and laboratories. As such, the NSF Policy Office has always had a fundamentally important job, as well as a challenging one: policies must be clear but at times highly flexible. As Feldman animatedly articulated, "Heavens, even in a directorate, you will find things fundamentally different between each of the divisions . . . We try to write [policies] flexibly. We try to make sure that those things that we really do care about are clearly articulated in the policy document. We've moved to a lot more system enforcement of the things that we really consider near and dear."[28]

The difference between policy and practice with the external side of FastLane was abundantly clear with regard to NSF's policy on keeping annual and final reporting up-to-date and enforcing deadlines. The policy that all PIs must submit final project reports before consideration for future funding for themselves and any co-PIs has long been a clearly stated and formal NSF policy, yet it was inconsistently enforced. FastLane enforced it in absolute terms. Similarly, FastLane now electronically enforces the required inclusion of a postdoc mentoring plan (for any project including a postdoc). While standard, consistent enforcement of these NSF policies is clearly sensible, the rule that reports must be in is not necessarily a popular one with a co-PI who can't submit his or her own new project until a colleague, the PI on the prior one, sends in a final report. As with the external side to FastLane, not everyone appreciated eJacket's effective role in policy enforcement.

Some program officers believed that eJacket may have been used to overstep, that it possibly moved into the territory of dictating policy. As an program officer in the Division of Ocean Sciences expressed the view, "Before eJacket . . . each group within NSF did things fairly differently, in part because of their internal structure. So Ocean Science has an extra level of management that for example, Polar [Programs] doesn't . . . So we had an extra loop of review and concurrence. And it was pretty easy to do. You could just make the process your own because it was just a piece of paper . . . Once we started doing things electronically . . . things become more policy-driven, in that an electronic system can drive policy . . . Some people were like, 'Cool, this is an electronic system, this is so much better; I don't have to have paper all over.' And others were like, 'Oh my God, it's tyranny! They're making us do things in ways we don't want to do.' "[29]

Such perspectives were not limited to program officers but also evident with administrative support managers and staff, many of whom are involved with using eJacket and also are witnesses to the impact of the system on a range of different program officers. The senior administrator for the Directorate for Biology stated, "eJacket, more than anything, changed the whole social environment . . . it changed what was really a more dynamic matrix process into something that was very linear . . . we . . . established an Office of Emerging Frontiers, . . . purposely designed [so] that all four of our division directors would co-manage it. And eJacket could not accommodate it. And because of the long list of upgrades required, and lack of funds, and the fact that this was directorate specific, it is always at the bottom of the list . . . We learned to work around it, and I'm afraid it didn't foster this interdisciplinarity . . . a directorate without walls; in fact, [it] put more walls up."[30]

Despite best efforts by DIS and the Policy Office, it was not possible to provide and express in electronic form a perfectly clear articulation of all the policies and practices in each directorate and its constituent divisions. And, even had it been, DIS developers faced difficulties in balancing the differences among directorates not to mention building in the needed flexibility to create a single working system. While Feldman fully understood that writing policy is "sometimes like art . . . it's got to be carefully crafted,"[31] there were limits to assembling exact formal knowledge of the way everything is done throughout

NSF, not to mention the need to find appropriate trade-offs to implement one overarching computing system.

Both the policy and technical challenges to getting eJacket implemented for declinations were considerable, but this was nothing compared to the far more complex and ongoing processes needed to go paperless for successful awards. As such, DIS was able to launch and refine a preliminary or beta version of eJacket for declinations in the 2003 to 2004 timeframe and achieve a fairly stable system that rid NSF of paper jackets for the unfunded 70 percent of proposals. The complexity to develop and implement policy in a software system to finalize an award decision and implement an awards management electronic infrastructure was immense and, due to funding constraints, significantly delayed.

As DIS director Andrea Norris recalled, "eJacket was supposed to be a pathfinder . . . the bridge between the old paper world and the new . . . [It] was only supposed to have to have a relatively short window of time [for development]. Because of funding, it stretched out to much longer."[32] Despite early hopes to launch a system that could handle both declinations and awards by the first years of the new century, functionality for awards was not available for a half dozen years after functionality existed for declinations. Only in late 2009 was eJacket set up for handling awards management, and the system was not in use throughout the foundation until the following year. It spawned a mixed reaction from its prime user base of NSF program officers.

Many program officers welcomed eJacket with open arms. They stressed the great advantage of being free of the paper clutter and ending the lost time searching for, retrieving, and shuffling jackets. They emphasized the advantage of a system to organize needed actions fully and neatly in eJacket's individualized "my work" tab. Some indicated how important the system had been (since its partial launch for declinations in 2003) to handling the agency's fast-rising proposal volume and that it would be difficult to imagine handling the ever-growing load without the efficiencies made possible in part by eJacket. One great advantage, according to some program officers, was that they were no longer tied to their offices to do their work. While paper jackets could never be taken from the NSF building, program officers could log in to eJacket anywhere around the world with a secure web connection. They could get work done during slow times at a professional conference, during halftime at a kid's sporting event, or at home. For the academicians who make up NSF's program officer workforce, this new flexibility to do work anywhere has been, and continues to be, important. As one Division of Physics program officer put it, "A lot of times you want to be somewhere else; you want to telework; and you don't have to physically be here to do your work. I think that's huge."[33]

Past program managers since promoted to higher levels of management were aware of different advantages with eJacket relative to recordkeeping and auditing. As the division director for Social and Economic Sciences stated, eJacket "has made it extremely transparent. Recordkeeping is now 99 percent accurate with program officers not having to be accountants, literally, to where they now have all the data with regard to reviews, with regard to budgets, with regard to recommendations, with regard to abstracts, in a single place that is available to them and is available to everyone in management. In the old system . . . [it] was much more likely to cause errors."[34] With regard to official record retention, DIS, the Policy Office, and the Office of Administrative Services recently coordinated with the National Archives and Records Administration (NARA), and received its approval for eJacket's electronic files to be the official NARA record of proposal management for NSF.[35]

Despite many at NSF seeing advantages to eJacket, a substantial minority of program officers perceived eJacket differently—either as a mixture of good and bad compared to the paper-based days or even as a tool that ran counter to advancing the work of NSF. Frequently enough, different program officers viewed the same functionality in completely different ways. Some praised eJacket for its approximation of replicating paper jackets in electronic form, while others argued that it was a lost opportunity not to start fresh and design a system more in line with evolving web standards for usability.[36] Regarding more specific functionality, most believe eJacket made it easier to do co-reviews between programs and to collaborate with their fellow program officers—to make a proposal available to another program for co-review is a few simple clicks. On the other hand, others pointed out that this could reduce face-time, discussion, and interaction between program officers. A program officer in Mathematical Sciences observed that a class of their proposals in differential geometry readily connects to certain programs in the Physics division. Rather than go down the hall, the program officers can now interact entirely electronically.[37]

Along with reservations about eJacket dictating policy, some significant critiques centered on the system's overall design, slow response times, and inordinate number of "clicks" required to get routine tasks done, as well as a handful of particular problematic modules. A meaningful minority of program officers believed eJacket has led to inefficiencies and even slowed down certain steps of their work. "FastLane in general is an enormous asset. It's just that some of us don't see any timesaving aspects to the all-electronic awards . . . the paper jackets were just as fast," noted one. Another program officer gave an example of how a cross-directorate co-fund, because of a difference in "review style," resulted in the proposal being "returned," and "it starts from scratch again, basically the whole process." So the cascade of time-consuming clicks had to be repeated. In this instance, eJacket "has become, at least at the award end, more cumbersome."[38]

Program officers also had mixed responses concerning DIS training for eJacket. Many program officers are short-term rotators rather than permanent NSF employees, a trend that has accelerated in recent years. As such, each year a substantial number of new program officers join the foundation for two- or three-year cycles. With eJacket being the central nervous system for their work, they need to become adept with it fully and quickly. For some longtime NSF program officers, the many years working with paper jackets and forms inclined them to be resistant to change since they had already mastered quickly reviewing paper jackets. These longtime program officers wanted an electronic system that worked well from the start, not one with glitches, such as missing codes for standard items. They also wanted quality training in how *they* would need to use the system. For some, training sessions fell far short of their expectations. One longtime program officer and deputy director of the Engineering Education and Centers division described the training as "ghastly." "The person teaching the course didn't know enough about how we actually use eJacket . . . He didn't know how we actually use it to do work. So it was horrendous." Her biggest complaint was that the designers and developers did not grasp the type of knowledge required to use the system as a program officer, that the eJacket development team did not find out the "needs of the user."[39]

To be sure, program officers from various divisions of NSF served on eJacket advisory committees. But the results of wide-ranging and not-always-consistent feedback made it difficult for the developers to complete the system on a tight budget. Scattered feedback in a committee meeting room did not necessarily convey how program officers do their daily work with the demanding flow of proposals. As a program officer in the Geosciences directorate who served on DIS eJacket committees related, "I tried from my early days to convince the developer that they needed to have closer collaboration with program officers to understand how they handle proposals and . . . was always sort of told that there was not enough resources to have some developers spend some time in the divisions . . . to shadow people to understand how many do you do on a given day, how many reviewers are involved, how many proposals and how many panelists . . . [The absence of shadowing] made it a little more bumpy than it could have been."[40]

By most accounts, eJacket has steadily improved since its rollout in full (including award processing) in late 2009, but some problems remained. The program director for Gravitational Physics emphasized that she is very good with computer systems but ran into problems with the budget module and, despite persistence, could not get a budget through the system in 2010 for one awarded grant. At the same time, she praised its functionality for generating review requests and reminder letters. Overall, she captured the widespread perspective of program officers in praising some attributes of the system while critiquing others, through it all remaining hopeful. Her larger message was that despite its problems, "it's continually improving."[41] Even though eJacket appears to have had many more glitches and challenges than the external side to FastLane ever did, it was the case (as examined in chapters 4 and 5) that such troublesome challenges as server capacity and PDF creation produced strongly voiced critiques from users (PIs and SRO staff). And it may be the case that interviewing PIs and SRO staff after many had experienced more serious challenges with Grants.gov may have led them to remember selectively and to minimize the early hurdles with FastLane.

Like early proponents of FastLane in the research community, many program officers at NSF already strongly praise eJacket, and its critics may modify some of their assessments as the system is gradually refined. If positive lessons for computer system design and development might be distilled from the experiences of the NSF users of eJacket, these three top the list: (1) have system designers shadow numerous program officers throughout the organization to gain understanding of user perspectives and their work requirements; (2) recognize the obligation for equal funding for internally facing systems and external ones; and (3) create training with explicit attention to how users will actually use the system. All of these guiding principles help achieve active cooperation and buy-in from users. They were followed with (the externally facing) FastLane by forming the external advisory committee (and keeping the same participants together and advising for several years) along with the highly knowledgeable outreach team for FastLane. Moving forward, the legacy code of the existing eJacket system, and its design structure that sought to re-create certain elements of paper jackets, will likely limit major changes. At the same time, through minor modifications as funds allow, eJacket likely will become increasingly stable, and possibly even a bit stale, for several years until its successor, by necessity, is developed largely from scratch.

eJacket and human resources changes and challenges

There was one other major complaint some program officers had about eJacket: that it shifted to themselves routine clerical work formerly performed by support staff. This shift in work clearly occurred and was noted by many program officers and support staff we interviewed.[42] Program officers often came to the foundation expecting their work to be almost entirely intellectual, with perhaps a bit of higher level administrative management. Longtime program officers sometimes did not appreciate the change in their work to include more mundane administrative tasks, while recently hired program officers (both permanent and rotators) frequently were surprised when they arrived and understood the actual mix of their responsibilities. Long-serving and new program officers alike sometimes saw it as inappropriate or inefficient for them to be responsible for basic administrative functions. The introduction of eJacket for declinations and later for awards greatly accelerated and expanded trends partially evident with the widespread use of earlier computer, software, and networking technologies at NSF.

Prior to the Erich Bloch era beginning in the mid-1980s, most program officers at NSF and most support staff did not have a computer at their desks. Typewriters or scattered word-processing workstations were used to generate correspondence with reviewers, proposal applicants, and others. Much of this was done by support staff and virtually every program officer had his or her own personal secretary. Taking dictation, fast and accurate typing, and careful filing and organization were necessary support-staff skills in the pre-PC era at NSF. With sizable numbers of new program officers (both for permanent and rotating positions) arriving each year, the secretaries and other support staff played a critical role in teaching new arrivals about the methods, bureaucracy, record-keeping requirements, and even the politics at the foundation. There were considerable demographic differences between program officers and support staff. The former were predominantly early-middle-age-to-older white male PhDs.[43] The latter were largely working class African American females, whose education often was limited to a high school diploma or, more rarely, a two-year associate's degree.

By the late 1980s, most program officers began to type out and send some proportion of their own correspondence from their own PCs. This expanded with the proliferation of email in the early 1990s—internally and between program officers and the research community. The launch of PARS in the mid-1990s also meant program officers generally did more administrative tasks. In the distant past, program officers might maintain a list of handwritten names of researchers. They would ask support staff to obtain addresses and write standard (or dictated) correspondence asking researchers for an individual review, or perhaps they would provide a long list of names to send correspondence asking researchers to serve as review panelists. PARS had a database and a form-letter library, so program officers sometimes came to do this entirely themselves. Even with these work assignment shifts, a major clerical task for support staff remained: filing and maintaining the mountains of paper, including tens of thousands new paper jackets each year. In 2004, eJacket largely shifted this responsibility (in electronic form) to program officers for 70 percent of proposals (the declinations), and in 2010 it shifted this work to them for 100 percent of proposals.

So eJacket meant that fewer support staff were needed overall. At the same time, support staff who could contribute in new ways to offset the increased load on program officers became increasingly valuable. As former NSF chief information officer George Strawn put it, all of a sudden "there was a bigger learning curve for support staff."[44] Some support staff were adaptable to develop new skills, as well as science and engineering reference skills (to identify potential reviewers based on the subject of grant proposals). For the latter, mostly the new job category of "science assistants"—rather than the former positions of secretaries or program assistants—was needed. Some science assistants were graduate students in scientific or technical fields; others had a BS from a prestigious school or had already completed a MS degree.

Middle-level NSF administrative support staff who had completed some higher education (or held BA or BS degrees) often thrived in the new environment. Sonya Mallinoff came to NSF with much of her college degree completed in 1971 as a congressional assistant in the Office of Legislative and Public Affairs. She later participated in a program for upward mobility for women within NSF (an affirmative action initiative to address the paucity of women in managerial ranks), completed her BS, and moved into one of the research organizations as a junior budget analyst. In this computerized environment, she was highly successful and now holds a senior administrative appointment in the Directorate for Biology.[45]

Some of those coming to NSF at one of the lower entry-level positions, such as clerk typist Daphne Marshall, adjusted well and now help with adding and checking codes in eJacket, finding reviewers on PARS, assisting with travel, coordinating arrangements for panels, and other computer and administrative tasks.[46] Losing some prior work roles to program officers, and having to learn new ones, was sometimes a blessing but initially created some anxiety.[47] As Marshall put it, "The anxiety is something that you put up some defenses about because you don't think that you're going to be able to adjust to it . . . But the information they provided us in hindsight was just what we needed."[48] Others coming to NSF with just a high school education for a strictly clerical job never adjusted well, and it has been very difficult for them, as well challenging for support staff managers and program officers trying to find ways for them to continue to contribute to the work of the foundation.[49]

Since NSF's top management was committed to not laying off any staff, extensive training and reorganization efforts occurred with computerization and, in particular, with the deployment of eJacket. Just prior to the launch of eJacket for declinations, NSF initiated a four-year study of the foundation's business processes that focused heavily on analyzing information technology and human resources. NSF hired management consulting firm Booz Allen Hamilton to carry out the study, which was overseen within the foundation by a veteran MBA-trained manager, Joseph Burt, who had worked within many different offices, including the Office of the Director. Booz Allen Hamilton proposed a major new organizational design, but NSF was unable to implement much of the critical human resources side of this design. As Burt explained, "We actually posited the idea that all of the program staff for a directorate could come under the control of a manager at the directorate level so that it would become a whole lot easier to juggle support staff from division to division, within a directorate, as needed to meet work flow. That one never got off the ground. There were a whole lot feathers ruffled over that. If you just look at it just purely from a workflow-management standpoint, it made a lot of sense. But there were a lot of cultural and other issues that came into play there."[50]

Some reorganization took place and opportunities for promotion were created by implementing findings of the Booz Allen study. Mallinoff's advancement to become a senior administrator, overseeing the human resources support staff of the Biology directorate, is a testament to this. At the same time, the study's wider recommendations for an extensive tiered management system, advanced training infrastructure, and flexible human resources allocation environment could not be accomplished. As Burt articulated, "Probably the single big thing that we never really got implemented the way that we wanted was we created a whole new three-level structure for program support staff. That was going to provide an opportunity for some folks . . . to have opportunities for real advancement and also opportunities for much more extensive training and development. And it was going to, over time, help us bring in fresh talent at a higher performance level. But that never, that didn't materialize . . . which was really the heart and soul of the model . . . That we never got to implement that is just a crying shame."[51]

The human resources side to eJacket is a stern reminder of the critical interrelationships between computing systems and employee management, established work cultures, and the great challenges and opportunities that accompany the implementation of transformative computing technologies.

User-driven innovation

Program officer critics of eJacket found the system not only wanting in design and operation but also lacking in certain functionality. They wanted new automation tools to help ease their heavy workload. NSF's DIS often receives suggestions, but their budget is pushed to the limit keeping FastLane and other legacy systems up and running. They simply do not have the resources to be developing additional tools—even ones that make a great deal of sense in lightening workload for program officers. In a sense, DIS has borne a great deal of the burden in the effort to develop technology (and oversee contractors) to help the foundation review ever more proposals without significant growth in staff size.

Compliance checking is an early and particularly tedious part of the proposal evaluation process. Noncompliant proposals are technically ineligible and thus should not go through formal ad hoc or panel review. Compliance includes page limits, margins, and other basics, as well as necessary justification statements—if there is a postdoc in the research design, there must be a postdoc mentoring plan; investigators and other researchers with more than two months per year on a project must provide an explanatory statement. In all, there are dozens of points of compliance, each of which takes time to check and can be subject to human error. Some support staff help, but ultimately program officers are responsible, and many spend their valuable time doing this checking themselves or serving as a second set of eyes after staff has looked over a proposal.

Paul Morris came to NSF in 2009 as a temporary program officer in the Atmosphere and Geospace Sciences division, filling in for a program officer on detail. He is a PhD scientist from the United Kingdom who moved to the United States when his wife, also a scientist, began working at NSF. Morris had worked previously as a telecommunications consultant and had developed programming and software engineering skills. Once at NSF, with the responsibility to manage the evaluation for a large program, he was faced with the burden of compliance checking. He soon thought there must be a better way. He informed DIS of his intentions as a courtesy (they were supportive), and in his spare moments in 2010, over the course of roughly 10 months, he programmed an 8,000-line front-end (to eJacket) "robot" compliance-checking tool in the AutoIt scripting language. AutoIt was well suited for Morris's endeavor; it is a BASIC-like freeware language with accessible syntax designed for automating GUI and general scripting. The tool checks a large number of the standard requirements, such as proposal length, type size, margins, required forms given certain project criteria, matching budget figures, and other basic compliance elements. It converts PDF proposal components to text to make this compliance checking possible and then generates a report for results. It takes about a minute for each proposal.[52]

Morris developed the script to make his life easier and to be helpful to other program officers who cared to use it. He refined the working prototype on roughly a weekly basis based on comments from users, issued new releases about once monthly, and it has now grown to approximately 10,000 lines of code. Word of Morris's compliance-checking tool spread quickly, and soon dozens of program officers were using it, many on their entire set of proposals. Morris kept lines of communication open with DIS, and recently DIS has taken it over to support it, placing contractor developers on the task and including it on the "laboratory" portion of its Research.gov site. With formal change requests to be considered and approved, DIS and its contractors have incorporated standard measures that have slowed releases of new iterations. With responsibility passing to DIS, Morris's involvement has waned: "It slightly takes the fun out of it, that's for sure, but I certainly understand why they [DIS] have done that as well." The tool ran for years on two old computers: one running compliance checks on even-numbered proposals, the other on odd-numbered proposals.[53]

The response to Morris's compliance-checking tool has been favorable within the program officer and support staff communities. In 2011 Morris estimated that roughly 30 percent of proposals coming into NSF were run through the compliance checker. It works in the background, so program officers or staff can do other work while checking is being done; many input a list just before they leave in the evening, and the compliance reports are waiting for them the next morning. Most program officers we asked were aware of it, and most had tried it. Many have made it part of their regular routine. Other program officers have tried it but think it generates too many false positives to be useful. Those who use it regularly agree that there are commonly false positives, though they think this limits the scope of where they need to concentrate and meaningfully saves time.[54]

While Morris's compliance-checking tool stands out for its widespread use and significant efficiencies, program officers have a much longer history of innovation for potential and realized systems, dating all the way back to EXPRESS, if not earlier. Program officers, as well as other staff, throughout the foundation frequently possess great creativity and technical talent, including programming and software engineering skills. Given tight DIS budgets, a substantial backlog to evaluate change requests, and small staff, it is no surprise that some initial NSF computing system innovation comes from outside DIS, particularly in computing-focused research directorates such as Computer and Information Science and Engineering. There, division director Michael Pisani employed a graduate student to program a system to help intelligently automate proposal clustering and finding panel homes for "orphan proposals." Regarding this effort, George Strawn, who has also been involved with interagency collaboration including Research.gov, commented, "The true promise of that is yet to come, but we've seen enough pilots that I'm willing to bet on it."[55]

Research.gov

FastLane, as a grant-submission and grants-management system, was truly pathbreaking. Though it continues to be greatly appreciated in the scientific research community, it lacks common web design interfaces of recent years, and it is expensive to support. Likewise, eJacket, despite only being rolled out in full form in 2010, has a decade of legacy code and is also a difficult and costly set of tools to maintain.

The rationale for standardization of electronic research administration for the federal government is strong, despite it being extremely difficult to implement due to unique elements of the different grant-awarding agencies and departments. Large state block grants from the Department of Transportation or Department of Energy are completely different from a NSF research proposal. Even NSF and NIH research proposals are far different in what, by policy, they need and require, as well as in post-award grants management. In hindsight, many of the difficulties with the attempt at a standard proposal system, Grants.gov, are entirely understandable.

After some initial glitches in the first years (with server crashes and PDF challenges) FastLane possibly led to some complacency for NSF; it had a system that worked well for the foundation and its research community. The possibility that NSF might have to require proposal submission through Grants.gov drove home the point that interagency cooperation had to mean more than regular interagency meetings and an openness to share ideas (which NSF had long done); it required active development of systems and tools that could and would be used jointly by different parts of the federal government. In the words of George Strawn, "Research.gov began several years ago with the admonition, don't duplicate the Grants.gov business."[56]

Research.gov was barely an idea when we began this study of FastLane in 2007. As such it was not part of our original research plan. As we are writing, Research.gov is the new host for the required annual and final project reporting to NSF. It is likely that Research.gov will gradually replace FastLane entirely. Additionally, it will offer new tools to the research community and agency staff that NSF, in Strawn's words, "couldn't have . . . dreamed about in the paper days" when FastLane was first conceived.[57]

Several years after the launch of Grants.gov, the federal government started the Grants Management Line of Business (GMLB). Like its brethren, the Financial Management Line of Business and the Human Resources Line of Business, GMLB was a multi-agency initiative focused on finding opportunities that could make the federal government more effective and efficient with its procedures and information technologies. One emphasis in GMLB from its earliest days forward was supporting agencies interested in providing shared services. NSF put in a proposal to GMLB for Research.gov, which was one of the chosen initiatives. Research.gov, a NSF-led effort partnering with NASA and several other agencies, currently is focused on providing common services to the external research community interacting with federal funding agencies.[58]

The National Science Foundation's motivation to develop Research.gov was in many respects a response to the Bush administration's Office of Electronic Government and Information Technology. In 2004 the head of this office, Karen Evans, had issued an explicit statement that agencies either will cross-service other agencies or will be cross-serviced by other agencies. On external-facing grant submission and grants management, NSF leaders believed they had long had superior technology with FastLane. But FastLane, as a collection of modules, was designed specifically for NSF and was an aging legacy system. In the words of NSF's Erika Rissi, a manager within the agency's Division of Grants and Agreements helping to oversee Research.gov, the initiative grew out of NSF believing it needed to "come up with a good strategy so we can continue . . . [to] provide these services in-house because that's what we think our community expects . . . In some ways it's akin to what Grants.gov was *trying* to do, but it's focused more broadly than 'find and apply,' and it's tailored specifically to the needs of the research community . . . One of the big lessons learned out of Grants.gov is that if you design to the lowest common denominator, the more sophisticated users are not going to have any patience with your solution and they will not use it."[59]

Research.gov's current focus is on external-facing services, but the broader strategy with the initiative is "holistic and ever-evolving." So while the first priorities are on such areas as proposal submission, post-award management, reporting (to NSF by PIs), and interaction with reviewers, in the longer term NSF conceives the initiative to include modernizing platforms for the agency's internally facing business services related to grants, such as review functions, creating and managing panels, internal reporting, and other features—in other words, also modernizing eJacket, replacing functions where it makes sense, and complementing it with new services as well.[60]

Some of the challenges moving forward with Research.gov—especially funding and developing what soon enough become legacy systems—are similar to those with FastLane and eJacket, but other challenges are even more pronounced given today's rapid pace of change in information technology. Like other agencies, NSF was never intended to engage directly in information-technology development, and accordingly such technology development typically gets contracted out. Funds are limited and technologies are path dependent. NSF managers pick technologies (both standard and custom) based on available data and perceived needs at the time but then can be locked in for years, missing out on incompatible but potentially useful new computing functionality. As Rissi stated, "In no way, shape, or form is FastLane cutting-edge technology. It is not. And it was at the time [of its launch]. And people still like it because it's pretty simple and it works, and at the end of the day that's the most important thing from the perspective of the researcher. But from an internal perspective, it's hard to maintain, . . . it can't integrate with other tools, . . . [and] it causes all sorts of problems."[61]

In having insider knowledge of and access to pioneers at the forefront of browser technology (such as NCSA's Mosaic), NSF was able to insightfully and skillfully develop FastLane (especially its external-facing side) as a set of innovative tools that truly pioneered and changed the world of grant submission and grants management. Visionaries within NSF—particularly Constance McLindon, Fred Wendling, Carolyn Miller, and Craig Robinson—profoundly influenced these achievements. With eJacket (the internal-facing side of FastLane), the reality of rapidly evolving web-based tools, specialized functional needs, and funding challenges led to long delays. And though it ultimately resulted in a working system that by most accounts aided efficiency, in many respects it was old by the time it was fully implemented. With Research.gov, NSF is taking a pioneering role once again; this time, however, that role is less about creating anew than about skillfully deciding how existing information technology can be selected, modified, deployed, and managed to fit the needs of the foundation, its agency partners, and the research community. As Rissi speculated, "What's going to happen is eventually we're going to modernize FastLane in Research.gov, and then it will stay there, and it will stagnate. And 10, 15, 20 years later there will be another big push to innovate, and we'll catch up to where the entire world is at that point in time."[62]

7

Legacies, Lessons, and Prospects

From its earliest conception in the 1990s, FastLane was intended to be thoroughly interactive, a step toward full-blown, multimedia, colleague-to-colleague interactivity. There were at that time inspiring visions for using computers to create rich and compelling communication spaces. So even if FastLane took practical form with the Mosaic/Netscape web browser—the software application that defined the 1.0 version of the World Wide Web with *static* webpages that seem old-fashioned today (see figure 4)—it was aiming at something beyond the one-way transmittal of images and text. In several respects, FastLane's creators aspired to push the existing state-of-the-art technology in the direction of Web 2.0 with its *dynamic* content and user interactivity.

FastLane came of age when notions of computer-supported cooperative work captivated the computing world and much of the research community, as described in chapter 2. Computer networks beginning in the 1960s had simple facilities for transmitting text, numbers, and other forms of fixed data. Proprietary networks had already linked computer users inside their respective corporate realms—IBM, Control Data, DEC, and others—while basic text-only commands, datasets, and simple email exchanges flowed across the ARPANET from the 1970s and in the next decade also across the Internet, conceived as a universal network of computer networks. The idea of using computer technology to transmit complex scientific *ideas*, however, was something new. Creating a means for interacting with full multimedia capability, including images, text, graphics, sound, and video, attracted the attention of visionary computer scientists at Carnegie Mellon and the University of Michigan who worked on

FastLane's influential predecessor, EXPRES. Absent anything like the World Wide Web, EXPRES posited a translation model where diversely formatted documents might be centrally stored and then called up and modified using diverse computer platforms, an arrangement that anticipated Google's "software as a service" office suite known as Google Docs. Dan Atkins, one of the PIs for the EXPRES project, readily admitted that their research vision of multimedia interactivity departed substantially from the more prosaic goal of automating the submission of NSF proposals; indeed, the two went in divergent or "orthogonal" directions.[1]

Fast forward from the late 1980s until 2003, after FastLane was successfully launched. Atkins was then lead author for the blue ribbon NSF report *Revolutionizing Science and Engineering through Cyberinfrastructure*, a far-reaching blueprint for revamping and expanding the foundation's computing efforts. "A vast opportunity exists for creating new research environments based upon cyberinfrastructure," it argued enticingly and persuasively. Cyberinfrastructure, a newly conceptualized layer of computer-based infrastructure, stood midway between the general-purpose internet computers, storage, and communication and the discipline- or domain-specific software and applications, consisting of "enabling hardware, algorithms, software, communications, institutions, and personnel."

Three years later, in 2006, Atkins was named to head NSF's newly created Office of Cyberinfrastructure, which was given a hefty budget of $127 million. NSF clearly appreciated Atkins work on EXPRES and his advocacy of cyberinfrastructure. As the NSF press release put it, EXPRES "laid the foundation" for FastLane. Both were clear expressions of Atkins's research interest in the architecture of distributed knowledge communities. One of the exciting concepts in the computer-supported cooperative work community was "collaboratories," a new type of digital laboratory where researchers could interact with colleagues by sharing data, information, and computational resources, without sharing the same physical location. Atkins affirmed in our interview with him that FastLane was an early instance of cyberinfrastructure.[2]

Lessons from listening

After conducting 400 in-person interviews across the country and collecting an additional 400 interviews through our online interview platform, we have an immense dataset on what designers intended or hoped for with FastLane, what users wanted and expected from FastLane, and how the process of implementing a nationwide distributed computer system worked. We also heard about many features of information systems that users hoped to have in future incarnations of FastLane, such as the emerging Research.gov. This concluding chapter identifies lessons that can be learned from our extended case study.

Curiously enough, in our age when social media platforms like Facebook and Twitter continually redesign themselves in the never-ending chase for novelty and the all-important advertising dollar, FastLane's freedom from the latest "bells and whistles" was frequently lauded. Interviewees noted that the basic front-end features and general usability of FastLane had changed little since 2000, apart from the much-appreciated online PDF conversion. This measured pace of change was welcome especially to busy researchers, since it helps them manage the high-stakes process of preparing and submitting proposals with consistency. Not even the very most active PI researchers use FastLane with the weekly or even daily frequency that is common enough among sponsored projects staff and some NSF staff. Only the most active PIs, who put in multiple proposals each year, might have occasion to develop such a degree of familiarity with FastLane, and it is exactly these active PIs who often have administrative support from their laboratory, center, or department to handle the mechanics of submitting proposals. These active PIs were insulated from the intricacies of Grants.gov as well. Sponsored projects staff reported steep learning curves and substantial frustration when learning the Grants.gov system. (Our discussion of the state-of-the-art in grants management comes later in this chapter.)

One important lesson for user-interface and system designers, then, is to *plan updates to accommodate how frequently the system is accessed by a typical user—rather than by experts.* If a department-level accounting specialist has daily interaction with a computerized accounting system, it is not an undue burden to plan frequent updates. But, by contrast, if many PIs submit a proposal with FastLane only once every few years, assuming the happy scenario that they successfully secure a three-year grant to support their research, it is a significant burden to face sets of major changes with each successive round of proposal submission. Whether intended or not, one consequence of rapid changes in a computer system (whether FastLane or another) is to tilt the field in favor of the most frequent users of the system and those users who have the greatest institutional assistance in using it. Conversely, the consequence of rapid changes is to shift stiff learning curves and extra work for mastering updates to the researchers and the institutions that have fewer resources for specialized assistance in developing proposals and submitting them to the agency. Generally, with FastLane's measured pace of change coupled with extensive assistance provided by NSF staff, we believe that these differential effects on users were intentionally modest.

A pressing and perennial question with computers is how and when they cause change in economic and institutional arrangements. Our research revealed a complicated pattern, somewhat different from the popular enthusiasm that "computers change the world" directly and automatically. Instead of the economic and institutional changes being driven by computer technology alone, we found that the dynamics of institutions—their values, policies, culture, and even budgeting cycles and priorities—can have far-reaching effects in what types of changes were sought from computers and which effects were brought about through them. Universities, research management, and the NSF itself are the key institutions that shaped the emergence, deployment, and consequences of FastLane.

During the earliest days, when FastLane was still on the drawing board, a tension emerged between having the new system closely mimic the existing paper proposal versus using its development to thoroughly overhaul, or even reengineer, the basic NSF proposal. While there were advocates within NSF of both the "paper paradigm" and the "reengineered proposal," there was no point in time (so far as we can tell) when a formal decision in favor of one was made. Implementation inclined toward some directions more than others: this is exactly what you might expect from a decentralized development process shaped by strong core values. FastLane largely expressed NSF's existing core values (including sanctity of merit review, confidentiality, and security) and was emphatically not used as a mechanism to impose changes on those values.[3]

FastLane's proposal submission module generally followed the paper paradigm and continued the agency's research-management practices that had emerged in the postwar decades, as chapter 2 relates. As manager Carolyn Miller noted, the watchword within the FastLane development team was "'Well, just make it look like the paper' . . . We just took the paper form and put it on the screen."[4] Many interviewees praised the welcome convenience that FastLane automatically created entries for the table of contents and summed up annual figures into a cumulative project budget. But the basic 15-page paper proposal that existed prior to FastLane is pretty much the same document that exists electronically in FastLane today. For years after FastLane's launch in fall 2000, and until the eJacket system was fully implemented, the first step for a submitted electronic proposal when it arrived at NSF was to be laser printed; from then on it was handled as a traditional paper document.

The persistent paper paradigm for NSF proposals is not universally admired. Several physical and biological scientists among our interviewees noted that FastLane has not offered features that have become customary in online journal articles. Moving images and interactive graphics are common in conference presentations, and naturally the journals publishing research work have tried to keep pace. A popular method is using the Adobe PDF format to embed interactive three-dimensional graphical images into what looks, to a computer network, like a standard text-and-image PDF document.[5] NSF, through FastLane, has generally resisted related features such as extensive hyperlinking to "off-page" data or additional documents as well as the use of interactive multimedia document formats. These formats are needed to support video images, simulations, and user-manipulated graphics.

Recalling some unimplemented features of FastLane suggests what a more fundamentally "reengineered" proposal system might have looked like. An early vision for FastLane floated the idea of radical transparency evidently to lessen the insistent pressure on program officers, who faced a barrage of phone calls and emails from PIs anxious to hear the latest word on their proposal's status and updates about their likelihood of funding. An early version, publicized in *Science* magazine, outwardly offered to make reviews available to PIs in something like real time. The *Science* article quoted with approval the observations of a geochemist, "You need practically all 'excellents' [in proposal ranking] to have a chance for funding. Ideally, FastLane would post each rating as soon as it was sent in, and if I saw any 'fairs' I'd know it was time to start working on a revision." Just three weeks later, as if to counter this suggestion of sweeping transparency, NSF deputy director Anne Peterson sternly emphasized that "merit review outcomes will still be communicated to proposers personally by program officers."[6] Even today, researchers usually learn the fate of their proposals directly from their program officer, with the proposal decision and status (along with reviews) available in FastLane a few days later.

A second lesson of our extended case study is that *computerization can recreate and even entrench existing organizational processes and procedures*. Focusing on the basic form and function of an NSF research proposal, the impact of FastLane was rather modest—certainly compared with NSF's thoroughgoing and constructive responses to congressional pressures for accountability. (The more dramatic impacts of FastLane are noted below.) One might suggest that the classic NSF proposal, whether paper or electronic, is an "evocative object" that expresses and embodies the NSF community's values.[7] It strikes a delicate balance between a researcher's desire for ample space to describe his or her ideas, a reviewer's hope for concise and readable proposals, and the program officer's need to manage an immense flow of information. Building a proposal around the basic 15-page project description had obvious economic ramifications when proposers first paid for 10, 15, or even 20 paper copies of each proposal, depending on the program they were submitting to, and then paid express postage to ship the package off to NSF.[8] NSF imposed a 15-page limit in an effort to streamline proposal length, moderate the expectations that proposers might feel obliged to fulfill, and limit the time burden imposed on proposal reviewers. (A full proposal, when adding the required front matter and back matter, including budgets, justification, facilities, and other required sections is typically 60 or more pages in length.) Clearly, with electronic PDF documents transmitted over the Internet, one can easily imagine research proposals (if there were otherwise no length limits and any slight advantage to greater proposal length) evolving such length and complexity that many might be the 350 pages achieved in some massive multi-institution EPSCoR proposals.[9]

The same concerns remain today about time burdens, efficiency, and equity, even though the direct economic cost of printing proposals has subsided. Several interviewees remarked that their academic unit either actively discouraged or even outright forbid their printing out of proposals received for review, suggesting that the economics of paper has not disappeared. For many years, NSF has strictly limited anything that might add additional pages to the project description directly or indirectly through extensive links to external images, data, or other files. While the out-of-pocket economic costs of long proposals have eased, there remain good reasons for the 15-page paradigm that relate to proposer equity and reviewer burden. NSF now faces a challenge in holding on to these existing community values (and strictly length-limited proposals) while embracing new and emerging community expectations about complex graphics and simulations (and unavoidably relaxing those limits). This new environment may require a new proposal paradigm that is equally an evocative object expressing the NSF community's evolving values.

Other elements of FastLane, including the Interactive Panel System, suggest a third lesson: *computerization can result in fundamental changes in processes and procedures if intentions to do so align with institutional resources*. When we speak of

"FastLane" we should remember that two of its more important and impressive elements were focused in large measure on NSF itself. From the prototype described in the Weber report forward, NSF managers understood that significant savings in time, effort, and staffing could be achieved only if the agency's internal processes were automated along with electronic proposal submissions. The Interactive Panel System and the eJacket system were, respectively, the computerized systems for peer or merit review and for managing the evaluation of proposals within NSF (as chapter 6 describes). Both of these systems resulted in major workflow changes in how NSF conducts routine business but with differing funding and timeliness. A telling contrast is that while FastLane and the Interactive Panel System had significant "outward" reach into the research community and, accordingly, could be more easily and amply funded, eJacket was much more difficult to fund since it was deemed to be "inward" oriented. (The agency's internal processes creating significant external results in the form of extramural research grants did not change the funding policy.) Each of these processes were time- and labor-intensive.

The Interactive Panel System was a mainstay for the research community's peer review of proposals, or merit review, as it became known. During mid-1990s, NSF evaluated its "merit review with peer evaluation" system and decided to foreground the term "merit review" over "peer review." According to NSF historian Marc Rothenberg, as early as 1986, "the term 'peer review' was properly a restrictive term referring to the evaluation of the technical aspect of the proposal. However, for more and more federally funded research, 'technical excellence' was, in the words of the [1986] committee, 'a necessary but not fully sufficient criterion for research funding.' Acknowledging that the NSF (as well as other federal agencies) was using a wide range of nontechnical criteria as part of the decision-making process, the committee suggested that the term 'merit review' more accurately described the NSF selection process."[10]

Sanctity of peer or merit review is one of NSF's cardinal values, and FastLane was designed to preserve and sustain this value—not merely to automate a labor-intensive process. "In the merit review area—the *process* of managing panels [and] the *process* of managing ad hoc or mail reviews—is very labor intensive," noted one NSF manager. The paper-based era was also hugely resource intensive, especially after the proposals had arrived at NSF. Untold staff hours were required to sort and distribute proposals within NSF and to mail them out to reviewers. For conducting a review, each panelist received a box of two dozen or more proposals that NSF support staff had painstakingly assembled and mailed.

Early efforts to provide NSF laptops for panel members were promising but expensive, especially for directorates with very large proposal loads such as Small Business Innovation Research or Education, which might organize 30 or more panels at nearby hotel rooms with high-priced Internet connections.[11]

As its name suggests, the Interactive Panel System was designed to facilitate electronically mediated exchange between panelist-researchers. The basic setup inherited from the paper days was a two- or three-day meeting among the small group of panelists to discuss and evaluate the set of proposals and to make recommendations on funding to the program officer. Craig Robinson, an atypical NSF project manager who had a prior research career as an astrophysicist, became a strong advocate of creating an interactive panel system "not just simply emulating a simple paper process, which is what a lot of the work that had been done to that point focused on—but actually how you change workflow structures." Paper-based proposals and panel reviews created a mountain of paperwork (panelists brought their boxes of proposals to NSF for the panel meeting and had the responsibility to discuss each paper proposal while there), so it was no great leap of insight to capture the panel's deliberations on laptop computers and then route the electronic panel summaries, via the program officer, straight to the original proposer. Connecting laptops, Internet connections, and electronic access to proposals and external reviews, along with instant-messaging among the panelists, created a new interactive working environment. With the panel system, thought Robinson, "the panel summaries—the quality—started to get better because people were starting to interact. Instead of one panelist writing it and everyone agreeing to it, they started interacting with one another." As one experienced program officer put it, with the new workflows, computer-centered environment, and expanded panel responsibilities, "it took a while for the panels to get comfortable."[12]

The constraints on NSF's spending money on its "internal" functions dramatically slowed the development and rollout of eJacket. Even though NSF managers had recognized in the earliest planning documents the desirability of automating its internal processes in parallel with external ones (see chapter 2), development projects focused on the agency's internal processes were difficult to fund. So despite the same attractive vision of interactivity in the air—and despite the same forceful and enthusiastic backer (Craig Robinson) involved with both the Interactive Panel System and eJacket—the development of eJacket took seven long years. Serious development on eJacket in its present web-based form began early in 2003, and within two years "the main components were completed."[13] Not until 2010, however, was eJacket fully functional for awards. During this interval, with the unceasing flood of electronic proposals arriving at the agency, program officers improvised with mixed or hybrid paper-and-electronic workflow systems to get their work done. It is undeniable that a speedier development of eJacket would have saved time and aggravation, and most likely money as well. The slow development of eJacket, owing greatly to the difficulties of finding funding for an "inside" project, is one instance of an institutional mechanism slowing down the pace of computer development.

Clearly, internal automation of workflow processes at NSF had economic and institutional benefits. In our research, we located one early study that circulated widely within NSF and was cited frequently as compelling evidence for the time savings and economic benefits. In "Cost Comparison of Panel Review with and without FastLane," an analyst in the Office of Science and Technology Infrastructure assembled comparative statistics on one particular award's all-electronic program solicitation, proposal processing, and panel review. Significant costs savings occurred since, in addition to cutting out physical printing, "all uses of postal services were eliminated." This study found significant *time savings* at each step of the process—program solicitation (preparing and disseminating a public document), 83 percent time savings; internal proposal processing (including setting up panel review), 50 percent; and time from panel review to the issuance of awards, a range of 42 to 64 percent. *Cost savings* for these same steps were equally impressive (61, 49, and 39 percent, respectively). Overall, in this study FastLane trimmed 6 to 10 weeks from the overall proposal process and cut costs by slightly more than half (51 percent), saving $39.57 per proposal. This study might not have been exactly representative but, especially to FastLane's supporters, it made an undeniable positive case.[14]

Less distinctly recognized are the economic and personal costs that may accompany automation. Within computer science, the classic question of "what *can* be automated" has been reformulated by some human-centered computing researchers to focus attention on the inherent value-laden question "what *should* be automated?"[15] Even the term "automation" is a complex and fractious one; NSF director Erich Bloch, who had come from IBM and the computer industry, avoided using it: "I don't like the word 'automation.'"[16] In a roundabout way, the slow development of eJacket had the effect, whether explicitly intended or not, of hampering not only the benefits of automation but also the workplace disruptions of automation and consequently giving staff members time either to adjust to the new computer-centered workflow or to transition to other lines of work at NSF or elsewhere.

The impact of automation on staff, especially the lower-tier support staff who had specialized in handing the mountains of paper, was widely recognized within NSF, although some of our interviewees were reluctant to speak on the record about it. To describe this process, we identify specific perspectives on automation—but, in this section, to respect the concerns about this controversial topic, we do not give specific citations to the interviews.[17] Despite former director Bloch's unease with the term, "automation" was commonly used by our interviewees to describe the introduction of computer technology and its consequences for workflow and job descriptions. One interviewee's first job arriving as a student in the 1990s was as "office automation clerk." A dozen or more researchers, and a dozen or so NSF staff members, also referred explicitly to "automation" in describing NSF's efforts. One even invoked the classic image: "Automation always gives you the chance of doing stuff faster than you did before. Well that means you have to work harder to keep up with the faster that the new machines are working; think of Charlie Chaplin's *Modern Times* movie where he's trying to keep up with the assembly line."

Early on, the impact of FastLane and automation on jobs at NSF was an explicit concern. The view that "FastLane is going to put us all out of work internally . . . was a major concern" and "a constant concern" within NSF. "A lot of the program assistants thought that the program officers were going to take a lot of their tasks away . . . So [they] were concerned back then when [FastLane] first came onboard," stated one staff member. FastLane "certainly changed the balance of work," thought another staffer: "We probably stretched the support personnel more to learn the new automated procedures . . . There was a bigger learning curve for the program support staff. And, of course, more gripes." According to yet another, "We had a whole set of administrative staff who would help do a lot of that in the paper world, that the electronic world wouldn't accommodate anymore." It is clear that the number of lower-tier support staff dropped off during the 1990s as several of the labor-intensive jobs, including typing letters of decline and acceptance, reviewer requests, and routine correspondence, were reduced as more program officers began to use personal computers and, then, as FastLane reorganized the core activities within the agency. Throughout the process, it was repeatedly stated that no one would lose their job because of FastLane, owing to the possibility of reassignment, retirement, attrition, and adjustment. Yet "in reality fewer people eventually were needed to support the same functions that they did before." One consequence of this explicit staff concern about automaton was heightened attention to "better communications with people involved to explain what's going on."

Innovation at the edges

The Interactive Panel System, eJacket, and FastLane itself are organized innovations that came from the center of NSF. These innovations had high visibility, formal structure, and wide accountability. In our interviews with some NSF staff, we also uncovered many informal efforts at innovation. Some of these instances of innovation at the edges are scarcely known at NSF today and only to select agency insiders. To our knowledge only one of them was ever formally acknowledged by NSF, and there is almost no paper trail or record to document any of them. Such "user-driven" innovation has recently received attention in the innovation literature, owing to the work of Eric von Hippel and his colleagues, and deserves to be better understood and more widely recognized.[18] It is a helpful counterweight to the traditional model of innovation that is conceptualized as rationally planned and centrally directed research and development. This informal or user-driven mode of innovation is worth spotlighting here, for at least three reasons:

First, *informal innovation can bring entirely new products into existence*. In 2010 a temporary program officer in Geosciences, Paul Morris, developed a "compliance checker" that subsequently was formally adopted and spread widely throughout the agency (see chapter 6). It is a singular instance of informal innovation gaining official sanction. Every proposal the agency receives must be examined for compliance with such detailed requirements as margin sizes, page limits, and the necessary supplemental documents for such issues as postdoctoral mentoring, allowable staff time, and data management. Morris saw that many of these items could be scanned by a computer, and in his spare time over 10 months he wrote an 8,000-line program that automated these tedious tasks. Initially his own personal effort, literally "running on a box in a broom cupboard," the compliance checker is now an officially supported piece of software used by several hundred NSF program managers. Compliance checks are run overnight in large batches or individually in the background while program officers attend to other tasks. Much to their credit, the agency's computing specialists gave Morris great leeway while it was under development. "I just about got away with anything if I [said] this is a prototype and I'm not offering it as an NSF service," Morris recalls.[19]

Second, *informal innovation can "test the waters" with small-scale experiments.* Numerous efforts across the agency to use computers to automate and streamline the work of program officers predated the formal FastLane project. In chapter 2, we describe the efforts of Economics program officers to run their custom-made BASIC programs on an early Atari computer, connected to a borrowed television monitor, and their extended experiment with IBM personal computers running Lotus 1-2-3 spreadsheets. But their experiment to use email in the 1980s to receive incoming proposal reviews massively overloaded the NSF email server at the time; they had greater success using email for outbound communications to the reviewer and PI communities. Essentially, the Economics program officers were running a small-scale experiment in using computing to streamline and reorganize their work; a full-scale experiment would have been entirely unworkable. Peter Morris's compliance checker, too, began initially as a small-scale experiment for himself and then expanded outward and across the agency.

Third, *informal innovation can explore technical options and refine users' expectations.* While it might seem obvious today that an e-government system should simply adopt the technical systems used successfully in e-commerce, it is essential to remember that FastLane took form *before* there were accepted e-commerce models to follow; indeed, FastLane and Amazon were designed and developed during the same historical moment in the mid-1990s and using many of the same hardware and software (as chapter 3 explains). One can view the technology assessment efforts of Connie McLindon and Fred Wendling as an informal means of exploring varied technical options before there was a clear path forward. Even the prototype EXPRES project with its now cumbrous-sounding translation scheme was a useful way to explore options and refine expectations; and whatever shortcomings NSF experienced with the central translation model, today Google Docs uses something very similar with evident success.

Connecting to users

Software packages and computer systems are too often developed with only a vague notion of users in mind. FastLane was notable for its attentiveness to users within the agency and especially to the external users among the research community and sponsored project staffs. Here we make an appraisal of *how* FastLane's developers created meaningful interaction with FastLane's users. Software engineering as a field is concerned with the processes that shape the design, development, and implementation of complex computer systems. Identifying mechanisms to facilitate the "user-driven innovation" identified by Eric von Hippel may be increasingly important for successful companies and organizations. We can identify at least three different means the agency created to bring users into the process of design and development. As one of them put it, "One of the things that impressed me intensely about NSF was how seriously they listened to the user community that was involved in creating and developing FastLane."[20]

First, there were *complementary formal mechanisms to solicit user ideas and preferences* including internal NSF committees such as the FastLane Internal Review Committee and FastLane Internal Implementation Group, as well as external consultative groups that had agency and influence. While FIRCOM was a high-level internal NSF strategic committee "trying to direct the directions that FastLane was going in," FIIG was the internal implementation group set up "when we realized . . . we needed people who were going to tell us what was the best way to implement something." Carolyn Miller recalled that, for each major step of FastLane, "there was a huge publicity campaign that would have to be conducted."[21] Active volunteers were sought within the agency, especially as modules of eJacket became ready for testing and evaluation. NSF rotator Saran Twombly recalled, "We volunteered to be guinea pigs for the award processing [and] the budget."[22] Equally impressive were the agency's wide-ranging outreach to an external group of institutional "guinea pigs" (as Neal Lane termed them), the set of universities and colleges that were a test-bed for early versions of FastLane. Pamela Webb, quite possibly the first person to submit a live FastLane proposal, recalled her participation in a Federal Demonstration Partnership task force on electronic approval and routing systems, noting that "it was a natural transition from there to electronic routing of actual submission and development of submission of proposals."[23]

The 20 colleges and universities in this task force formed a tight working partnership with the developers at NSF. Sometimes, user testing is an empty exercise, but that was far from the case here. Pamela Webb, then at UCLA, recalls her "unusual . . . experience with the way these kinds of pilots worked . . . So they literally would take a screen or an area and prototype it, show it to us as the user group, and then get their comments, come back and show us another round of it before they even put out in a pilot phase. We would comment on it in a pilot mode, both with dead data, and then later with live data. And again, they would make the changes, and they made them timely—it was very responsive." NSF's Carolyn Miller called her up one day with a puzzle about the very first deadline for NSF graduate research fellowships. When she went home on Friday afternoon just 22 fellowship proposals were in the system, but when she returned to work on Monday morning, the day of the deadline, there were 480, and she thought certainly something was wrong. No one had imagined that proposals would flood in so close to the deadline![24]

Second, *on-site training of users before the system's launch* allowed extensive in-person handholding and spotlighted lingering problems. Even if PI researchers did not recall taking part in extensive special training to learn FastLane, the NSF staff interacted extensively with university and college sponsored-projects staff and, in turn, they often took the lead in helping researchers and faculty to learn the new system. At a dozen or more of our site visits, we learned of a single person (most often in the sponsored projects office) who was the local "guru" or resident expert in FastLane—and highly valued. University staff had the opportunity to meet with NSF trainers at their home institutions or at NSF regional meetings or at meetings of NCURA and SRA (the two principal organizations of sponsored research professionals), as well as at NSF headquarters. Once there were working versions of FastLane, NSF staff provided "actual FastLane sessions" at these meetings so that attendees could gain first-hand experience.[25] Again, NSF's funding structure encouraged this extensive travel and interaction since it was seen as an "external" purpose of assisting the research community. Interviewees also pointed out the complementary function of the NSF FastLane Help Desk, which provided in-person assistance over the telephone.

Third, an *ethos of proactive engagement with users* was internalized in the FastLane development effort. It was not only that FastLane was developed, as one senior NSF manager put it, with "very active involvement and engagement with the research community."[26] Even program officers got into the act. "We would go bug the computer development teams because we wanted functionality that wasn't there." [27] Craig Robinson personified FastLane's engagement with PI users—since he was one himself. Carolyn Miller recalled that even though the team had solicited input from researchers, his research background mattered: "You had that immediate feedback on whatever it was you were discussing and how that might impact the researchers. So having him there was wonderful."[28]

Value-laden design

Advocates of "value-laden design," such as philosopher Helen Nissenbaum and her colleagues, make the case that computing is a powerful means through which to embed or embody societal values. In their view, computing systems are *not* morally neutral or without consequence for society, nor are they mere instruments for doing useful work; just the opposite, computing systems are critical ways of "building in" explicit attention to values such as concern for privacy or informational freedom or individual autonomy or property rights, or conceivably other values like surveillance and control.[29] When Stewart Brand coined the slogan "information wants to be free," he was arguing idealistically in favor of informational freedom and against a generation of media companies that have tried to monetize it (and even though the statement is little remembered today, Brand simultaneously acknowledged that "information also wants to be expensive").[30]

Computer systems today are mostly developed with the hope of commercial success, defined as market share or the share price of a publicly traded company. It is often difficult to tell whether an individual company's marketplace success in building and selling effective software products or computing systems reflects its success in attentive value-laden designing or effective marketing or, just possibly, fortuitous timing. With FastLane we can make a detailed appraisal of how widely shared institutional values—such as confidentiality, security, and peer or merit review—were translated into, and indeed embedded in, the computer system. These are generally held values evident in talking with dozens of program officers up and down the halls of NSF. In addition, the central repository of NSF's institutional values is its Policy Office. Owing significantly to its energetic chief, Jean Feldman, the Policy Office interacted continually and constructively with the FastLane development process: from early designs through field training of sponsored-project research staff. In short, *institutional values articulated by NSF's policy unit were integral to system development*. As Feldman told us, simply but emphatically, "I'm certainly a logical person to speak to, as I have worked with my colleagues in the Division of Information Systems in designing electronic systems for both internal and external use that meet the policies and requirements of the National Science Foundation."[31]

Feldman came to the NSF in 1992 with a background in research policy from the Department of Defense. She had experience working in policy appointments with the research-oriented Air Force Office of Scientific Research and the Army Medical Research and Development Command. Her NSF appointment "wasn't a difficult challenge in terms of research policy because I had been doing that for a number of years already. But it was moving from a DoD to a civilian agency and that is very, very different." She began at NSF as a grants policy specialist then worked her way into the supervisory track and eventually became head of the Policy Office. She became engaged with the agency's discussion of electronic proposals in the mid-1990s when FastLane was in its early development.[32]

It is no surprise to find that Feldman's Policy Office was greatly respected by and had close ties with the NSF units charged with developing and sustaining FastLane. It "was critical to get her involvement . . . her support and communication," noted one key staff member. Andrea Norris, head of DIS, observed that "FastLane reflects community-centric needs but the policy's in the foundation. So there were close interrelationships with that." Norris saw that policy played a subtle role in articulating institutional values across the foundation as well as influencing the details of eJacket. In the paper-based days, "the directorates really had [not] come up with a common way of doing things," but during the development of eJacket, "policy [and] general counsel [were] heavily involved, as they should be." NSF's core institutional values articulated through the Policy Office and specified in "policy checks and balances" accordingly shaped such details as "capability for any of the email correspondence, notes, informal things, anything." The Policy Office's influence on eJacket extended so far into the system's details that one program manager thought there were "policy clicks" solely "to verify that a [NSF staffer] has actively sought that link."[33]

It is important to realize that values are not static requirements and that, even in the federal government, policy is not always a fixed entity. "The problems are not the technology; the problems are the policy, and getting the policy to be consistent with the technology," noted FastLane manager Carolyn Miller. Craig Robinson, as Miller's successor during the launch months in 2000 and subsequently as lead for the Interactive Panel System, was in a position to see the interaction of policy and technology. "So we had to be creative, we had to have flexibility, we had to have a policy office here that worked with us . . . to adapt policies to meet the changing requirement of our electronic version," he noted. Robinson, too, emphasized that the paper-based era permitted great latitude in procedures and some slippage in enforcement: "We had some complex, oddball programs that . . . didn't fit into the policies of the Foundation, but no one ever even knew it," until eJacket came along. Here he credits Feldman's Policy Office for sensible flexibility that brought even the "oddball programs" into line with the NSF-wide policies. Miller, too, appreciated the flexibility in easing the 30-day deadline for no-cost extensions and other small but consequential points.[34]

NSF values influenced FastLane's rollout far beyond the headquarters offices in suburban Washington, DC. During the extensive field training across the country, "it was really important to have someone from the Policy Office with us, because quite often, the questions were not technical, they were policy questions," remembers Miller. Feldman recalls significant input from university researchers when evaluating the issue of PI signatures, which in the paper-based days imposed a significant burden on PIs to collect the signatures from department chairs, deans, and vice presidents or provosts, as well as the practical nightmare of doing collaborative proposals with a similar set of signatures coming from another institution. "I can guarantee you that when we . . . eliminated PI signatures, it was [owing to] the input from the PIs," she noted.[35]

Grants management 2.0

Our final set of observations on "best practices" in research management makes visible the immense diversity of institutional practices that we observed in talking with top-level university administrators and the supervisors of sponsored research offices. A key point is this: there is no single all-embracing "model" that can be recommended as "best" for all universities. American research universities embody substantial diversity in organizational styles, in their institutional culture, and in the size, focus, and research strategy of key laboratories or programs. For some departments, institutes, and colleges, NSF is the most significant funding agency and so, for these, it might make some sense to focus on "optimizing" their institutional arrangements for NSF. For most universities and all universities of large size, however, NSF is one funding agency among many. Universities with prominent research-oriented medical schools are more closely aligned with the National Institutes of Health while many universities with prominent physical-science and engineering programs may be more connected with the research offices in the Department of Defense, Department of Energy, or NASA.

It is an axiom for many in research administration that *investments in research support pay for themselves*, owing to expanded research activity and greater indirect-cost recovery, but it is not always clear how the investments might best be structured. We observed in our fieldwork both "centralized" and "decentralized" research administration. Universities practicing a decentralized model often transferred research support or even SRO staff from a central university-wide office to a college-level, or sometimes even a department-level, office. PIs at these universities generally praised the greater individualized attention they received during proposal preparation from having one staff member who had substantial experience and expertise in their specific area. A dramatic case of specialization is the University of Washington's computer science department, which has five grants-oriented staff that work directly with individual groups of faculty; other institutions ranging from the University of Massachusetts Amherst to the University of Alaska Fairbanks reported similar department-level setups.

One liability with a decentralized research-support model is that for faculty members, proposal deadlines are not spread evenly throughout the year. NSF once had the practice of accepting research proposals on any day of the year, but the agency's shift to general "targets" and specific "deadlines" means that time-specific pressures have become intense. A centralized model of research support, by providing a "pool" of highly qualified grants specialists, can spread out the inevitable time crunch of discipline- or program-specific deadlines over a larger staff. One thing is clear: prominent departments in well-funded fields, as well as at universities on the rise, such as Arizona State, are investing more rather than less in research support.

Staff training is invisible but essential. From the vantage point of PI researchers, NSF's extensive effort in training was, for all intents and purposes, invisible; few faculty members recall any specific training in learning to use FastLane. Nonetheless, there were informal networks or "pockets of excellence" that formed on many campuses: when one person learned how to submit proposals on FastLane, a necessary skill for research faculty, that person helped his or her colleagues and peers. Many universities were lucky in that their FastLane expert was in the sponsored research office (or in a campus-wide EPSCoR office) and so could lend assistance across the campus. It was crucial to have "at least one person who [was] the true guru in the [SRO] office," recalled one veteran.[36] FastLane "gurus" sprouted up in many diverse locations: at one university it was the director of undergraduate research, another was in ocean science, and yet another in the history department.[37]

Awareness of system-to-system software is now required for upper-level research administrators. When we began our research FastLane and Grants.gov were our main topics of interest, and we quickly learned of other agency-specific grants portals such as NASA's NSPIRES system.[38] Research.gov was gaining ground as the strategic successor to FastLane, as our interview with NSF's former CIO George Strawn makes clear. FastLane itself gained a reprieve when NSF suspended the officially mandated use of Grants.gov during the 2009 American Recovery and Reinvestment Act (ARRA) owing to concerns that use of Grants.gov would slow down NSF's processing of an additional $3 billion in research funding.[39]

In the wake of the ARRA "stimulus funding," many universities began to explore different options for submitting proposals through the Grants.gov portal. The "system-to-system" model directly connects a university's computing system to the receiving agency's (see chapter 5). The University of Hawaii during our site visit in 2011 was adopting one such software package, Kuali Coeus, an open-source product. A chief motivation was that, as the SRO director put it, it was "very cumbersome . . . to use so many different electronic submitting systems" across the federal government. At Hawaii the software package was tailored to work with Grants.gov but not FastLane or NSPIRES. Other universities have chosen commercial software packages like Cayuse, Click Commerce, SmartGrant (GAMS), InfoEd, Key Solutions, and others, which also tend toward integration with Grants.gov. It seems likely that this "system-to-system" model will be attractive to university grants administrators seeking to deal with the baffling computing complexity of diverse federal agencies while simultaneously facing substantial pressure for efficiency and cost-cutting. These are developments that everyone in the research community needs to be aware of.[40]

We can now step back, take a long-term perspective, and consider how computers alter our institutions and, in so doing, shape our society. It is not a simple story, as we hope readers can by now appreciate. Because NSF was created in 1950, it had developed a mature set of institutional values and policies that long predated the introduction of computing. Indeed, FastLane was developed by and introduced into a complex and dynamic institution. In this set of concluding remarks, we first offer a "snapshot"-oriented set of comments; next discuss some direct data from our interviews; and then distill some generalizations about computing and automation from these observations.

In a "snapshot" view, typically comparing "before" and "after" the introduction of a technology, it is deceptively easy to infer that changes in technology are causal. To return to the bicycle and automobile (chapter 1), you might infer that bicycling in the 1890s led women to discard the heavy garments and restricted

Victorian-era gender roles or that American suburbs resulted from automobiles. And you might imagine, in the words of the *Economist* magazine, that concerning the globalization of trade, "containerisation changed everything." Even an ordinarily cautious historian, in a bold essay on "Moore's Law and Technological Determinism," provides a listing of typical "before" and "after" classroom technologies (manual typewriters, blackboards, and chalk versus cell phones and Microsoft PowerPoint), concluding, simply but categorically, "I have made my point: Moore's law is at work."[41] Such snapshots incline toward a technology-centric causality.

But let us recall that snapshots are not fully rounded narratives with the capacity for embracing multiple voices and diverse perspectives. Snapshots fail to notice the flow of history. For FastLane in particular, and computerization and automation generally, technologies are typically introduced into dynamic societies and interact with ongoing institutional, cultural, and economic changes. It takes some effort to see this. One reviewer of this volume offered the speculation that FastLane resulted in an undue emphasis on the "broader impacts" of research. In like measure, one might—neatly but fallaciously—appeal to the FastLane computer system to explain far-reaching changes at NSF: the agency's greater attention to "broader impacts" of research, increased emphasis on interdisciplinary research, such high-profile efforts as the 1990s human genome project, or even wariness about funding potentially controversial lines of research. There was a lot in the air during the FastLane years.

In this instance, such technology-centered speculations are wide of the mark. By contrast, we would rather emphasize ongoing developments within the sciences toward interdisciplinary research, greater awareness of and interest in lines of research that engage the wider world, and the powerful presence of Congress—all of which substantially preceded FastLane's development and implementation. Congressional scrutiny following the Man as a Course of Study controversy in the 1970s (during the Watergate years) prompted NSF to refine peer review and streamline the proposal-reviewing process, which set the stage for computerizing the agency in succeeding decades (see chapter 2). So, of course, MACOS did not cause FastLane, nor did FastLane cause broader impacts, interdisciplinarity, and congressional scrutiny.[42]

To explore these questions, we asked our interviewees about the impact of FastLane on their research and the research system. Most PIs treated FastLane as a useful means to communicate their science to NSF and the review community. Some PIs and NSF staff praised FastLane for supporting and facilitating collaborative research involving PIs from more than one institution. Some interviewees felt that FastLane was overly committed to the paper paradigm and should be more receptive to multimedia, simulations, motion pictures, and other complex forms of presenting research data that are increasingly the norm in their scientific fields. Interviewees with extensive NSF experience in the paper days sometimes disapproved of the "box-like" nature of the reporting modules. With paper reporting, these PIs felt they could explain their results in a narrative style and holistic manner; with electronic reporting, they felt constrained to "fit" their research results into one of several boxes, such as Activities and Findings, Training and Development, Outreach Activities, several types of Publications, and five distinct types of Contributions. Proposals, before and after FastLane, remained a mix of open-ended narrative (the 15-page project description) and boxes requiring specific institutional and budgetary information. Several interviewees suggested that email, even more than FastLane, brought major changes in communication among researchers and with NSF staff.

NSF staff members are intensive uses of FastLane and eJacket (as chapter 6 explains), and they offered pertinent observations. NSF staff and university SRO staff often commented that FastLane helped manage significantly increased proposal volumes in an efficient manner. "We couldn't have dealt with the paper just because of the huge volume," thought one program director. "When you think about the volume we do today, it's just amazing to me that we are even still here," said one university research manager.[43] Procedural standardization was another concern. Several NSF staff indicated that whereas during the paper-based years there was considerable procedural variation across the agency, FastLane imposed a degree of standardization in reviewing, compliance checking, and decision timelines. FastLane also provided ready means to monitor individual program officers and to enforce NSF policies. To give a simple example, it was common with paper-based proposal reviews to split one's ranking between, say, "excellent" and "very good" (if the proposal were between the two), and the FastLane team struggled to accommodate such split ranking, as a database might normally accept just one value ("excellent" or "very good" or other defined ranking) per field. A tug of war ensued with NSF managers moving back and forth between permitting split ranking or endorsing single-value ranking. FastLane was the means to enforce this policy decision. More than that, FastLane was a centralizing tool that made visible the (otherwise hidden) cross-agency diversity and so in a way created a policy dilemma for NSF managers to address.

Computers change society, and it is because social, cultural, and institutional values can be embedded in them that the details of computing matter so deeply. FastLane is one instance where explicit attention to users resulted in a computer system that, by and large, was congruent with the institutional values of NSF and the research community. NSF designed and developed FastLane to express its cardinal values (see chapter 1). With the rise of commercial off-the-shelf computing systems and integrated enterprise resource planning (ERP) packages, users of these systems typically lack access to the companies designing the hardware and writing the software that they will use. "University administrators may be drawn to the organizational features of ERP systems but often fail to remember that they have been designed to centralize corporate business processes," states one journalist.[44] It remains for us to invent effective means to influence the computing developments that run our institutions and, ultimately, that structure our economies and societies.

Appendix A

University Site Visits (interviews and type)

	People interviewed	EPSCoR[a]	HBCU[b]
Arizona State University	16		
Boise State University	11	Y	
Florida A&M University	6		Y
Florida State University	11		
Howard University	11		Y
Jackson State University	8	Y	Y
New York University	11		
North Carolina A&T State University	7		Y
North Dakota State University	12	Y	
Purdue University	15		
Santa Clara University	3		
South Dakota State University	7	Y	
Stanford University	15		
University of Puerto Rico	10	Y	
University California, Berkeley	18		
University of Alaska Fairbanks	16	Y	
University of Arizona	6		
University of Hawaii at Hilo	5	Y	
University of Hawaii at Manoa	19	Y	
University of Massachusetts Amherst	17		
University of Minnesota	11		
University of Montana	15	Y	
University of New Hampshire	21	Y	
University of North Carolina at Greensboro	5		
University of Notre Dame	13		
University of South Dakota	4	Y	
University of Texas at Austin	15		
University of Virgin Islands	6	Y	Y
University of Washington	19		

Table A.1 continued

	People interviewed	EPSCoR[a]	HBCU[b]
Total universities visited	29	12	5
Total university people interviewed	333		
NSF interviews (10 nonrecorded, 69 recorded)[c]	79		
Total of in-person interviews	412		

[a]Experimental Program to Stimulate Competitive Research (see www.nsf.gov/epscor).

[b]Historically black colleges and universities (see tinyurl.com/mnn8zox).

[c]In November and December 2006, Misa conducted nonrecorded interviews at NSF with Dan Atkins, Edward Hackett, Fred Kronz, Daniel Newlon, Robert O'Conner, John Perhonis, Tom Weber, and Fred Wendling as well as phone interviews with Connie McLindon and NSF rotator Bruce Seely.

Appendix B

Interview Summary Statistics

During the course of our research, we did in-person interviews with 413 people; an additional 400 people completed our online interview. The following tables tabulate interviewees by state, gender, discipline, and years from PhD. The figures do not always total 413 or 400 since, for instance, we could identify an online respondent's gender, discipline, state, or years from PhD only if that person explicitly told us. Sometimes, we omitted asking an in-person interviewee about their discipline or years from PhD. We compiled these numbers to inquire whether our online and in-person samples were reasonably comparable and whether we created an interview dataset that was broadly representative of the NSF research community.

Table B.1 Number of participants

	In person	Online
Alaska	15	—
Arizona	22	1
California	36	102
District of Columbia	11	1
Florida	17	1
Hawaii	27	—
Idaho	12	—
Illinois	—	1
Indiana	28	30
Maine	—	1
Maryland	1	—
Massachusetts	18	2
Minnesota	11	—
Mississippi	8	1
Montana	16	—
Nebraska	—	1
New Hampshire	20	—
New Jersey	—	2
New York	11	—
North Carolina	12	9
North Dakota	12	—

Table B.1 continued

	In person	Online
Ohio	—	3
Oklahoma	—	1
Pennsylvania	—	2
Puerto Rico	10	—
South Dakota	11	11
Tennessee	—	1
Texas	18	47
Utah	—	1
Virgin Islands	6	—
Virginia	72	1
Washington	19	76
West Virginia	—	1
United Kingdom	—	1
Total	413	297

Table B.2 Number of participants indicating gender

Gender	In person	Online
Female	168 (43.7)	76 (28.7)
Male	225 (57.3)	189 (71.3)
Total	393 (100.0)	265 (100.0)

Table B.3 Number of participants indicating discipline

Discipline	In person	Online
Biology (Medicine)	42 (19.0)	57 (20.4)
Chemistry	32 (14.5)	32 (11.5)
Computer science	15 (6.8)	14 (5.0)
Earth sciences	30 (13.6)	28 (10.0)
Engineering	34 (15.4)	57 (20.4)
Mathematics	17 (7.7)	18 (6.5)
Physics	26 (11.8)	29 (10.4)
Social sciences	25 (11.3)	44 (15.8)
Total	221 (100.0)	279 (100.0)

Table B.4 Participants' years from PhD

	In person (n = 186)	Online (n = 242)
Average	25.8	28.4
S.D.	10.0	11.0
Median	24.0	26.5
Maximum	59.0	59.0
Minimum	3.0	5.0

Notes

CHAPTER ONE: Managing Science

1. See ITU's "2014 ICT Facts and Figures" at tinyurl.com/mpaxbty.

2. See "E-commerce in China: The Alibaba Phenomenon," *Economist* (23 Mar. 2013) at tinyurl.com/cegaakl, and KPMG, "E-commerce in China: Driving a New Consumer Culture" (Jan. 2014) at tinyurl.com/lc2l9lx.

3. See David Kriesel, "Xerox Scanners/Photocopiers Randomly Alter Numbers in Scanned Documents" (2 Aug. 2013) at tinyurl.com/ouvlf96, and Ken Ericson, "Update on Scanning Issue: Software Patches to Come" (7 Aug. 2013) at tinyurl.com/l2mob9t.

4. Nick Wingfield, "Grand Theft Auto V Muscles Its Way to Sales Record," *New York Times* (18 Sept. 2013) at tinyurl.com/l2umv75; David Stout, "Grand Theft Auto V Slams into Controversy with Gratuitous Torture Scene," *Time* (19 Sept. 2013) at tinyurl.com /kn7jg7q. For philosopher Jean Baudrillard, such a simulacrum is not a faithful copy of the real, but rather, since it is a copy of something that did not originally exist, it might even become "hyper-real," or truth in its own right.

5. See Scott Shane, *Academic Entrepreneurship: University Spinoffs and Wealth Creation* (Cheltenham, UK: E. Elgar, 2004), 33–34, and Paul Andrews, "Profit without Honor," *Seattle Times* (5 Oct. 1997) at tinyurl.com/d49u5u6.

6. Robert Spector, *Amazon.com: Get Big Fast*, rev. ed. (New York: HarperBusiness, 2002), 50; "Ten Year Anniversary of First Transaction Submitted via FastLane," NSF press release (Mar. 2005), from NSF FastLane News Archive (CBI Project Files—FastLane): "On March 29, 1995, the first transaction (a review) was submitted via FastLane."

7. See "NSF FastLane Wins Nationwide Award for Best Use of Internet Technologies," NSF press release 96-076 (3 Dec. 1996), at tinyurl.com/ljg96pe.

8. During our research, several FastLane principals located for us dozens of relevant primary-source documents in their office files; these documents are now in the possession of the office of the NSF historian. NSF routinely collects two classes of documents: high-level National Science Board reports and proceedings and complete award "jackets," with all documents pertaining to a specific (successful) proposal. Many of the interesting developments in research funding, such as internal agency discussion on emerging interdisciplinary and cross-directorate programs, are not routinely archived. FastLane was in this latter category. In addition, we collected legacy versions of (e.g.) FastLane News Archive (1995–2008), UCLA's FastLane PowerPoint slides (1999–2000), NSF slides on Grants.gov, an early Compuware evaluation of FastLane (1996), and several NSF timelines (1940s–2000s); these are archived in CBI Project Files—FastLane.

9. For a time, NSF proposals were to migrate to Grants.gov (a rival grants-management computing platform used by NIH, USDA, and other agencies, described in chapter 5) even while NSF researchers still submitted grant reviews and grant reports through FastLane. Similarly, although NSF is moving its reporting to a new platform called Research.gov, Research.gov as recently as fall 2012 served as a portal that took users *back* to FastLane's reporting area. In FY2015 NSF's *total* budget was $7.3 billion, and it funded "24 percent of all federally supported basic research conducted by America's colleges and universities." In mathematics, computer science, and the social sciences, NSF is the principal federal source; see "About the National Science Foundation" at www.nsf.gov/about.

10. Bruno Latour, *Science in Action: How to Follow Scientists and Engineers through Society* (Cambridge, MA: Harvard University Press, 1987), 150, 245.

11. According to the most recent proposal guide from NSF, "Proposal and Award Policies and Procedures Guide," NSF 15-1 (26 Dec. 2014) at www.nsf.gov/pubs/policydocs /pappguide/nsf15001/index.jsp. The average funding rate for proposals across FY2009–12 was 25%, according to "FY2013 Agency Financial Report," figure 6, at www.nsf.gov /pubs/2014/nsf14002/tables.jsp. The 50 million pages is our estimate based on 15 copies of 40,000 paper proposals averaging 80 pages in length. A single EPSCoR proposal (discussed below), a complex multi-investigator award, can be 350 pages in length.

12. Seth Hettich and Michael J. Pazzani, "Mining for Proposal Reviewers: Lessons Learned at the National Science Foundation," in *Proceedings of the Twelfth ACM SIGKDD International Conference on Knowledge Discovery and Data Mining* (New York: ACM Press, 2006), 862–71.

13. Adobe PDF evolved from the company's established PostScript page markup language. A PDF prototype in 1991 was even named Interchange PostScript. PDF version 1.0 was announced in fall 1992, and the essential tool to create and view PDF files (Acrobat) was released in June 1993. For preparing a PDF for FastLane in the early years, you first needed to "print" a document to PostScript format and then use Adobe Distiller to create an NSF-ready PDF document. See "A History of PDF" (9 Aug. 2013) at www.prepressure .com/pdf/basics/history, which observes that "Acrobat Distiller was available in [a] personal [version] priced [initially] at $695."

14. Emi Ito Oral History, conducted by Jonathan S. Clemens, University of Minnesota (7 May 2012), 5. Emi Ito was a member of the agency-wide Committee on Equal Opportunities in Science and Engineering and Mathematics.

15. See J. Merton England, *A Patron for Pure Science: The National Science Foundation's Formative Years, 1945–1957* (Washington, DC: NSF, 1982); George T. Mazuzan, "The National Science Foundation: A Brief History," NSF 88-16 (July 1994) at tinyurl .com/695c6q; Toby A. Appel, *Shaping Biology: The National Science Foundation and American Biological Research, 1945–1975* (Baltimore: Johns Hopkins University Press, 2000); and Dian Olson Belanger, *Enabling American Innovation: Engineering and the National Science Foundation* (West Lafayette, IN: Purdue University Press, 1998). Studies of NSF grant making include George T. Mazuzan, "'Good Science Gets Funded . . .': The Historical Evolution of Grant Making at the National Science Foundation," *Science Communication* 14 (1992): 63–90, and Marc Rothenberg, "Making Judgments about Grant Proposals: A Brief History of the Merit Review Criteria at the National Science Foundation," *Technology & Innovation* 12, no. 3 (2010): 189–95. Our interviews give a wealth of insight into the *diversity* of research programs and practices within NSF and across the sample of 29 United States universities where we conducted our research. A differently designed longitudinal study would better address long-term *changes* in the character of federally funded science.

16. Chris Kanaracus, "Air Force Scraps Massive ERP Project after Racking up $1B in Costs," *Computerworld* (14 Nov. 2012) at tinyurl.com/c2v32lm; Dan Eggen and Griff Witte, "The FBI's Upgrade That Wasn't," *Washington Post* (18 Aug. 2006) at tinyurl.com/zkvq4; Michael D. Shear and Annie Lowrey, "In Tech Buying, U.S. Still Stuck in Last Century," *New York Times* (22 Dec. 2013) at tinyurl.com/lm6ldhg. The Oregon and Minnesota ACA website problems in 2014 are described in "Catalogue of Catastrophe" at *Why Do Projects Fail?* calleam.com/WTPF/?page_id=3.

17. See Thomas J. Misa, "Understanding 'How Computing Has Changed the World,'" *IEEE Annals of the History of Computing* 29, no. 4 (Oct.–Dec. 2007): 52–63.

18. Michael W. Whalen et al., "Your 'What' Is My 'How': Iteration and Hierarchy in System Design," *IEEE Software* 30, no. 2 (Mar.–Apr. 2013): 54–60, quote p. 54.

19. N. G. Leveson et al., "Requirements Specification for Process-Control Systems," *IEEE Transactions on Software Engineering* 20, no. 9 (1994): 684–707; Joaquín Nicolás and Ambrosio Toval, "On the Generation of Requirements Specifications from Software Engineering Models: A Systematic Literature Review," *Information and Software Technology* 51, no. 9 (Sept. 2009): 1291–307 at dx.doi.org/10.1016/j.infsof.2009.04.001; Barry W. Boehm, "Software Risk Management: Principles and Practices," *IEEE Software* 8, no. 1 (1991): 32–41; Barry W. Boehm, "A Spiral Model of Software Development and Enhancement," *IEEE Computer* 21, no. 5 (May 1988): 61–72.

20. Andrew Sears et al., "Defining an Agenda for Human-Centered Computing," *ACM SIGACCESS Accessibility and Computing* 91 (June 2008): 17–21 at doi.acm.org/10.1145/1394427.1394430.

21. Ahmed Seffah, Jan Gulliksen, and Michel C. Desmarais, eds., *Human-Centered Software Engineering: Integrating Usability in the Software Development Cycle* (Dordrecht: Springer, 2005).

22. Jonathan Grudin, "Three Faces of Human-Computer Interaction," *IEEE Annals of the History of Computing* 27 no. 4 (2005): 46–62, quote p. 53.

23. Patrick Brézillon, "Focusing on Context in Human-Centered Computing," *IEEE Intelligent Systems* 18, no. 3 (May 2003): 62–66; R. R. Hoffman, A. Roesler, and B. M. Moon, "What Is Design in the Context of Human-Centered Computing?" *IEEE Intelligent Systems* 19, no. 4 (July–Aug. 2004): 89–95; Matti Tedre, "What Should Be Automated?" *ACM Interactions* 15, no. 5 (2008): 47–49; Robert R. Hoffman and Stephen M. Fiore, "Perceptual (Re)learning: A Leverage Point for Human-Centered Computing," *IEEE Intelligent Systems* 22, no. 3 (May–June 2007): 79–83; A. Jaimes et al., "Human-Centered Computing: Toward a Human Revolution," *IEEE Computer* 40, no. 5 (2007): 30–34; Steven J. Landry, "Human Centered Design in the Air Traffic Control System," *Journal of Intelligent Manufacturing* 22, no. 1 (Feb. 2011): 65–72.

24. See Melvin E. Conway, "How Do Committees Invent?" *Datamation* 14, no. 5 (1968): 28–31, as well as the author's story (n.d.) at www.melconway.com/Home/Conways_Law.html.

25. See Tracy Kidder, "Flying Upside Down: The Hardy Boys and the Microkids Build a Computer," *Atlantic Monthly* (July 1981) at tinyurl.com/o8pdg4s.

26. Peter Zheutlin, *Around the World on Two Wheels: Annie Londonderry's Extraordinary Ride* (New York: Citadel, 2007), quote p. 27.

27. EPSCoR was intended "to strengthen research and education in science and engineering throughout the United States and to avoid undue concentration of such research and education"; see "Experimental Program to Stimulate Competitive Research (EPSCoR)" at www.nsf.gov/od/oia/programs/epscor/about.jsp. NSF, NASA, DOD, and other federal

agencies participate in EPSCoR activities. For recent assessments, see Paul Hill, "EPSCoR 2030: A report to the National Science Foundation" (Apr. 2012) at tinyurl.com/nylpp7g, and B. Zuckerman et al., "Assessment of the Defense Experimental Program to Stimulate Competitive Research (DEPSCoR): Final Report Volume I—Summary" (Institute for Defense Analyses, Oct. 2008) at www.ida.org/upload/stpi/pdfs/ida-d-3649-vol-1.pdf.

28. Katarina Prpic, "Science, the Public, and Social Elites: How the General Public, Scientists, Top Politicians, and Managers Perceive Science," *Public Understanding of Science* 20, no. 6 (2011): 733–50; Robert J. Brulle, Jason Carmichael, and J. Craig Jenkins, "Shifting Public Opinion on Climate Change: An Empirical Assessment of Factors Influencing Concern over Climate Change in the U.S., 2002–2010," *Climatic Change* 114 (2012): 169–88; Dan M. Kahan, Hank Jenkins-Smith, and Donald Braman, "Cultural Cognition of Scientific Consensus," *Journal of Risk Research* 14 (2011): 147–74.

29. National Science Board, *Science and Engineering Indicators 2014*, NSB 14-01 (Arlington, VA: National Science Foundation, 2014); workforce data from www.nsf.gov/statistics/seind14/index.cfm/chapter-3/c3s1.htm.

30. Atar Baer, Stefan Saroiu, and Laura A. Koutsky, "Obtaining Sensitive Data through the Web: An Example of Design and Methods," *Epidemiology* 13, no. 6 (2002): 640–45; Joyce J. Fitzpatrick and Kristen S. Montgomery, *Internet for Nursing Research* (New York: Springer, 2004); B. Janssen, "Web Data Collection in a Mixed Mode Approach: An Experiment," in *Proceedings of Q2006: European Conference on Quality in Survey Statistics* (2006) at www.statistics.gov.uk/events/q2006/downloads/W19_Janssen.doc; "Best Practices for Conducting Internet-Based Sexual Health Research," 2012 National STD Prevention Conference, cdc.confex.com/cdc/std2012/webprogram/Session12935.html; Sari L. Reisner et al., "Comparing In-Person and Online Survey Respondents in the U.S. National Transgender Discrimination Survey: Implications for Transgender Health Research," *LGBT Health* 1, no. 2 (June 2014): 98–106; Sven Alfonsson, Pernilla Maathz, and Timo Hursti, "Interformat Reliability of Digital Psychiatric Self-Report Questionnaires: A Systematic Review," *Journal of Medical Internet Research* 16, no. 12 (2014): e268 at doi.org/10.2196/jmir.3395.

31. See Arne Hessenbruch, "The Trials and Promise of a Web-History of Materials Science," in Karl Grandin, Nina Wormbs, and Sven Widmalm, eds., *The Science-Industry Nexus: History, Policy, Implications* (Sagamore Beach, MA: Science History Publications, 2004), 397–413; Arne Hessenbruch, "'The Mutt Historian': The Perils and Opportunities of Doing History of Science On-Line" (26 Nov. 2004) at tinyurl.com/lsxrvdu; Jeffrey W. Franklin et al., "Developing a Contact Strategy to Maximize Self-Administered Web Participation" (Research Triangle Park, NC: RTI International, 25 Apr. 2007) at tinyurl.com/lbjro5b; Don A. Dillman, *Mail and Internet Surveys: The Tailored Design Method* (New York: John Wiley, 2000); Don A. Dillman, Robert D. Tortora, and Dennis Bowker, "Principles for Constructing Web Surveys" (1999) at chnm.gmu.edu/digitalhistory/links/cached/chapter6/6_27_papers.htm; and Michael W. Ross et al., "Biases in Internet Sexual Health Samples: Comparison of an Internet Sexuality Survey and a National Sexual Health Survey in Sweden," *Social Science and Medicine* 61, no. 1 (2005): 245–52.

32. We conducted in-person interviews with 413 people (see appendices); an additional 400 people completed our on-line interview (a further 165 persons registered and answered some interview questions but did not complete the full four-page set of questions).

33. CBI oral histories are accessible at www.cbi.umn.edu/oh, including a selection of our interviews for this project with FastLane designers, developers, and managers. The public dataset is accessible at www.cbi.umn.edu/oh/fastlane.

CHAPTER TWO: **Origins of E-Government**

1. Harry Wilmot Buxton and Anthony Hyman, *Memoir of the Life and Labours of the Late Charles Babbage* (Cambridge, MA: MIT Press, 1988), quote p. 46 ("steam"); Charles Babbage, *Passages from the Life of a Philosopher* (London: Longman, Green and Co., 1864), quote p. 137 ("future course of science").

2. See John Agar, *The Government Machine: A Revolutionary History of the Computer* (Cambridge, MA: MIT Press, 2003); David M. Hart, *Forged Consensus: Science, Technology, and Economic Policy in the United States, 1921–1953* (Princeton, NJ: Princeton University Press, 1998); and Paul N. Edwards, *The Closed World: Computers and the Politics of Discourse in Cold War America* (Cambridge, MA: MIT Press, 1996).

3. Daniel J. Kevles, "The National Science Foundation and the Debate over Postwar Research Policy, 1942–1945: A Political Interpretation of 'Science, the Endless Frontier,'" *Isis* 68 (1977): 5–26; Larry Owens, "The Counterproductive Management of Science in the Second World War: Vannevar Bush and the Office of Scientific Research and Development," *Business History Review* 68 (1994): 515–76; George T. Mazuzan, "The National Science Foundation: A Brief History," NSF 88-16 (July 1994) at www.nsf.gov/pubs/stis1994/nsf8816/nsf8816.txt; Michael Dennis, "Reconstructing Sociotechnical Order: Vannevar Bush and U.S. Science Policy," in Sheila Jasanoff, ed., *States of Knowledge: The Co-Production of Science and Social Order* (New York: Routledge, 2004), 225–53.

4. See "The NSF Mission" at www.nsf.gov/nsf/nsfpubs/straplan/mission.htm.

5. Chandra Mukerji, *A Fragile Power: Scientists and the State* (Princeton, NJ: Princeton University Press, 1989); Jacob Darwin Hamblin, *Oceanographers and the Cold War: Disciples of Marine Science* (Seattle: University of Washington Press, 2005).

6. Two legislative histories of the rival Kilgore and Bush visions are David M. Hart, *Forged Consensus: Science, Technology, and Economic Policy in the United States, 1921–1953* (Princeton, NJ: Princeton University Press, 1998), 158–64; and Daniel Lee Kleinman, *Politics on the Endless Frontier: Postwar Research Policy in the United States* (Durham, NC: Duke University Press, 1995), 100–171.

7. Daniel J. Kevles, *The Physicists: The History of a Scientific Community in Modern America* (New York: Knopf, 1978), quote p. 295.

8. Daniel J. Kevles. "The National Science Foundation and the Debate over Postwar Research Policy, 1942–1945: A Political Interpretation of Science—The Endless Frontier," *Isis* 68, no. 1 (1977): 4–26. See Public Papers of Harry S Truman 1945–1953, 169 Memorandum of Disapproval of National Science Foundation Bill (6 Aug. 1947) at tinyurl.com/le5z6e8.

9. Vannevar Bush, *Science: The Endless Frontier* (Washington, DC: Government Printing Office, 1945), chapter 6: "The Means to the End," section "Five Fundamentals" at www.nsf.gov/od/lpa/nsf50/vbush1945.htm#ch6.3.

10. Russell A. Kirsch, "Computer Development at the National Bureau of Standards" (n.d.) at tinyurl.com/msn39yx; Kirsch, personal communication to Misa (Apr. 2009).

11. Larry Owens, "MIT and the Federal 'Angel': Academic R & D and the Federal-Private Cooperation before World War II," *Isis* 81 (1990): 188–213; Robert E. Kohler, "Science, Foundations, and American Universities in the 1920s," *Osiris* 3 (1987): 135–64.

12. Bush, *Science: The Endless Frontier* at www.nsf.gov/od/lpa/nsf50/vbush1945.htm#ch6.3.

13. President Truman rejected the first three candidates for NSF director before naming Alan Waterman (then at the Office of Naval Research) to the position, a post that he held through two six-year terms from 1950 to 1963.

14. Bush, *Science: The Endless Frontier* at www.nsf.gov/od/lpa/nsf50/vbush1945.htm #ch6.3.

15. Stuart Umpleby, "Heinz von Foerster and the Mansfield Amendment," *Cybernetics and Human Knowing* 10, nos. 3–4 (2003): 161–63.

16. A full study of NSF and its relations to the U.S. Congress is beyond the scope of this study, but NSF makes several formal reports to Congress, including the biennial *Science and Engineering Indicators* on science and engineering education and workforce, the biennial report of the Committee on Equal Opportunities in Science and Engineering, and financial and inspector general reports; see www.nsf.gov/about/congress/nsf-congress-reports.jsp.

17. George T. Mazuzan, "'Good Science Gets Funded . . . ': The Historical Evolution of Grant Making at the National Science Foundation," *Science Communication* 14 (1992): 63–90; Marc Rothenberg, "Making Judgments about Grant Proposals: A Brief History of the Merit Review Criteria at the National Science Foundation," *Technology & Innovation* 12, no. 3 (2010): 189–95; National Science Board, "National Science Foundation's Merit Review Criteria: Review and Revisions," NSB MR-11-22 (14 Dec. 2011).

18. Comptroller General's Report to the House Committee on Science and Technology, "Administration of the Science Education Project 'Man: A Course of Study' (MACOS), National Science Foundation" (Washington, DC, 1975); James J. Kirkpatrick, "Sex for the Eskimo Fifth Graders," *Lewiston (ME) Daily Sun* (1 Apr. 1975): quote p. 19 at tinyurl .com/a9skygn. For analysis, see Linda Symcox, *Whose History?: The Struggle for National Standards in American Classrooms* (New York: Teachers College Press, 2002), 19–24, and Ronald W. Evans, *The Tragedy of American School Reform* (New York: Palgrave Macmillan, 2011), 125–47.

19. Glenda T. Lappan, "Lessons from the Sputnik Era in Mathematics Education," Symposium on Reflecting on Sputnik (National Academy of Sciences, 1997), quotes (skeleton staff and worst political crisis) at www.nas.edu/sputnik/lappan3.htm. For a recent analysis, see Heather B. Gonzalez, "An Analysis of STEM Education Funding at the NSF: Trends and Policy Discussion" (Congressional Research Service, 2012) at tinyurl.com/o7m6vwc.

20. Robert Lovell, quoting a letter Proxmire wrote to the CBS program *60 Minutes* about space colonization in *L-5 News* 2, no. 11 (Nov. 1977): 1. In 2011, Senator Tom Coburn (R-OK) issued a critical Proxmire-style report entitled "The National Science Foundation: Under the Microscope." Senator Coburn's press releases were at www.coburn.senate .gov/public/index.cfm/pressreleases (accessed Jan. 2013, no longer available). NSF leaders pointed out that Senator Coburn's principal source was NSF's own internal evaluation.

21. Rep. John Conlan from Arizona, quoted in Mazuzan, "Good Science," quote p. 77.

22. Stephen Cole, Leonard Rubin, and Jonathan R. Cole, *Peer Review in the National Science Foundation: Prepared for the Committee on Science and Public Policy of the National Academy of Sciences*, 2 vols. (Washington, DC: National Academy of Sciences, 1978, 1981).

23. "On numerous occasions the program officer could state only good things about the proposal and explain that because of the close competition, selection had to be made not on the absolute scientific merit of the proposal itself but on such criteria as the relative emphasis of various scientific approaches or perhaps on geographic distribution," writes Mazuzan, "Good Science," p. 71, noting that these practices taxed NSF program officers' diplomatic skills and collegial relations.

24. In 1986 NSF adopted the term "merit review" to recognize the expanded criteria appropriate for "center-based activities, research groups, and shared facilities"; quoted in Committee on Science, Engineering, and Public Policy, *Major Award Decisionmaking at the National Science Foundation* (Washington, DC: National Academy Press, 1994), quote p. 62; see also Rothenberg, "Making Judgments about Grant Proposals."

25. Quotations from Erich Bloch, interview with NSF historian Marc Rothenberg (11 Feb. 2008): 3.

26. Gordon Bell, oral history interview with David K. Allison (Apr. 1995) at americanhistory.si.edu/comphist/bell.htm.

27. Vinton G. Cerf, "RFC 1167: Thoughts on the National Research and Education Network" (July 1990) at tools.ietf.org/html/rfc1167.

28. Bloch interview, 13–14. Connie McLindon, phone interview with Thomas J. Misa (4 Dec. 2006), recalled a long-range planning figure from 1984 of one personal computer for each four NSF staff.

29. Dan Newlon, interview with Thomas J. Misa, NSF (16 Nov. 2006).

30. R. Molesworth, "The Open Interchange of Electronic Documents (open document architecture [ODA])," *IEE Colloquium on Standards and Practices in Electronic Data Interchange* (21 May 1991): 1–6.

31. Frederic J. Wendling Oral History, conducted by Thomas J. Misa and Jeffrey R. Yost, NSF (19–20 Feb. 2008): 34.

32. Dan Atkins, interview with Thomas J. Misa, NSF (17 Nov. 2006). For an insiders' informed assessment of EXPRES, see Murray Aborn and Alvin I. Thaler, "On-Line Research-Support Systems," *Annals of the American Academy of Political and Social Science* 495 (Jan. 1988): 127–34. Aborn and Thaler drew attention to Starr Roxanne Hiltz and Murray Turoff's description (in 1978!) of "computerized conferencing in the grant-review process" (129) in their prescient *Network Nation: Human Communication via Computer* (Reading, MA: Addison-Wesley, 1978), 214–50. For perspective on computer-supported cooperative work as an emerging and multidisciplinary field, see Jonathan Grudin, "Computer-Supported Cooperative Work: History and Focus," *IEEE Computer* 27, no. 5 (1994): 19–26.

33. James H. Morris et al., "Andrew: A Distributed Personal Computing Environment," *Communications of the ACM* 29, no. 3 (Mar. 1986): 184–201; Craig Partridge, "The Technical Development of Internet Email," *IEEE Annals of the History of Computing* 30, no. 2 (2008): 3–29, at 20–21.

34. Jonathan Rosenberg et al., "Translating among Processable Multi-Media Document Formats Using ODA," in *Proceedings of the ACM Conference on Document Processing Systems (DOCPROCS '88)* (New York: ACM, 2000), 61–70, quotes pp. 61–62.

35. Rosenberg et al., "Translating among Processable Multi-Media Document Formats," 61–62; Jonathan Rosenberg et al., *Multi-Media Document Translation: ODA and the EXPRES Project* (New York: Springer, 1991); Gary M. Olson and Daniel E. Atkins, "Supporting Collaboration with Advanced Multimedia Electronic Mail: The NSF EXPRES Project," technical report 22 (Lansing: University of Michigan Cognitive Science and Machine Intelligence Laboratory, 1989).

36. Constance McLindon Oral History, conducted by Jeffrey R. Yost, Reston, VA (22 Sept. 2009): 8–9.

37. Stephen Lukasik Oral History, conducted by Judy E. O'Neill (17 Oct. 1991), Charles Babbage Institute OH232 at purl.umn.edu/107446.

38. McLindon Oral History, 4; McLindon, phone interview.

39. McLindon biography at CNRI website (4 Nov. 2002) at www.cnri.net (accessed 17 Nov. 2006). McLindon was director, Division of Information Systems (1980–88); director, Office of Information Systems (1988–92); and director, Information Technology and Resource Management (1992–96).

40. McLindon Oral History, 6, 8.

41. Wendling Oral History, 25–26.

42. Wendling Oral History, 106–7. The Unix-based Note system was installed and sup-

ported by Mike Morris. You can identify Note-era email addresses by the "note.nsf.gov"; see "NSF91122 Funding for Support of Women, Minority, and Disabled High School and Undergraduate Students as Engineering Research Assistants" (1991) at www.nsf.gov/pubs/stis1991/nsf91122/nsf91122.txt.

43. Ann Westine et al., "RFC 1168: Intermail and Commercial Mail Relay Services" (July 1990) at tools.ietf.org/html/rfc1168.

44. McLindon Oral History, 14, 15.

45. McLindon Oral History, 13; General Accounting Office, Report to the Chairman, Subcommittee on Investigations and Oversight, Committee on Science, Space and Technology, House of Representatives, "National Science Foundation: Planned Relocation to Arlington" (Washington, DC: GAO, Sept. 1992), quote p. 3.

46. Wendling Oral History, 10, 14.

47. Wendling Oral History, 106–7; "An Evolution from Software to Biology: Doug Walker Retiring CEO WRQ," *3000 NewsWire* (Feb. 2005) at tinyurl.com/mtxuna4.

48. Wendling Oral History, 20–24. On Ethernet at PARC, see Douglas K. Smith and Robert C. Alexander, *Fumbling the Future: How Xerox Invented, Then Ignored, the First Personal Computer* (New York: W. Morrow, 1988), 97–103, and Michael A. Hiltzik, *Dealers of Lightning: Xerox PARC and the Dawn of the Computer Age* (New York: HarperBusiness, 1999), 187–93, 363–64.

49. Wendling Oral History, 25.

50. For electronic federal research management in the mid-1990s, see Martin I. Kestenbaum and Ronald L. Straight, "Paperless Grants via the Internet," *Public Administration Review* 56, no. 1 (1996): 114–20. The authors found ready information on grant announcements and grant-management policies; paperless peer review was available through "NSF's test program FastLane"; electronic award notification was a long-term goal at NIH and available at NSF. Outside NSF, "few agencies will accept electronic submission of grant proposals" (119).

51. Wendling Oral History, 40–41.

CHAPTER THREE: **Developing a New System**

1. Gopher is anatomized in Farhad Anklesaria et al., "RFC 1436: The Internet Gopher Protocol (a Distributed Document Search and Retrieval Protocol)" (Mar. 1993) at tools.ietf.org/html/rfc1436.

2. Philip Frana, "Before the Web There Was Gopher," *IEEE Annals of the History of Computing* 26, no. 1 (2004): 20–41, quote p. 28.

3. "The Dot-Com Bubble Bursts," *New York Times* (24 Dec. 2000) at tinyurl.com/lb729v5.

4. For the collapsed stock valuations of 250 companies in the "Internet wasteland" a year after the March 2000 peak, see Anthony B. Perkins and Michael Perkins, *The Internet Bubble*, rev. ed. (New York: HarperBusiness, 2001), 289–96.

5. See Patricia B. Seybold, *Customers.com: How to Create a Profitable Business Strategy for the Internet and Beyond* (New York: Times Business, 1998), 157–72, quote p. 170.

6. Seybold, *Customers.com*, quote p. 170.

7. Vannevar Bush, "As We May Think," *Atlantic Monthly* (July 1945) at tinyurl.com/97k5nmv.

8. Wei, quoted in James Gillies and Robert Cailliau, *How the Web Was Born: The Story of the World Wide Web* (Oxford: Oxford University Press, 2000), quote p. 213.

9. T. Berners-Lee and Robert Cailliau, "WorldWideWeb: Proposal for a HyperText Project" (12 Nov. 1990) at www.w3.org/Proposal.html.

10. "Tim Berners-Lee: Confirming the Exact Location Where the Web Was Invented" (8 July 2010) at tinyurl.com/2uorgn2.

11. The base price for a NeXT computer, with a 25 MHz CPU and supplemental 25 MHz FPU, was $6,500 at the time, or perhaps double that figure (adjusted for CPI) today.

12. Gillies and Cailliau, *How the Web Was Born*, 210–17, 221–35.

13. "Andreessen Horowitz Announces $1.5 Billion Fund III," press release (31 Jan. 2012) at tinyurl.com/n394pd9.

14. "In the Beginning There Was NCSA Mosaic . . ." (n.d.) at ftp.ncsa.uiuc.edu/Mosaic/Windows/Archive/index.html (accessed Feb. 2013; no longer available).

15. Michael A. Cusumano and David B. Yoffie, *Competing on Internet Time: Lessons from Netscape and Its Battle with Microsoft* (New York: Free Press, 1998), 95. The 19 Feb. 1996 *Time* cover, "The Golden Geeks," is at tinyurl.com/mvl5v7y; Steve Lohr, "Spyglass, a Pioneer, Learns Hard Lessons about Microsoft," *New York Times* (2 Mar. 1998) at tinyurl.com/mg8lm8l. The subsequent history is somewhat tangled. The Netscape Company merged with AOL in a complex $10 billion transaction. Netscape Communicator (with web browser, html editor, email, news client, and address book) was spun off as a separate product and became the basis, through the Mozilla Foundation, for today's full-featured SeaMonkey and the popular web browser Firefox. Microsoft's Internet Explorer, through version 6 (1995–2006), acknowledged a Spyglass license from the original NCSA Mosaic.

16. Greg Newby Oral History, conducted by Thomas J. Misa, University of Alaska Fairbanks (28 Sept. 2011): 6. The NCSA renewal was PI Larry Smarr "The National Center for Supercomputing Applications" (award 8902829, $123,309,756) soon followed by "National Computational Science Alliance" (award 9619019, $249,066,609). The awards with Hardin as PI were "Visualization Tools in the MS-DOS Environment" (award 9014312, $60,000); "NCSA/Jackson State Collaborative Project: Visualization in the MS DOS Environment" (award 9115334, $76,702); "NCSA Hierarchical Data File Software Capitalization: National Distribution and Support" (award 9020884, $163,020); and "Exploratory Research and Initial Development of Software for the Analysis of Multiple Hybridization Images" (award 9016566, $50,000).

17. Fred Wendling, interview with Thomas J. Misa, NSF (15 Nov. 2006); Tom Weber, interview with Thomas J. Misa, NSF (16 Nov. 2006); George Strawn Oral History, conducted by Jeffrey R. Yost, NSF (10 Mar. 2010). PI Scott Lathrop, "Development, Enhancement, and Support of Internet-Based Software at the National Center for Supercomputing Applications" (award 9315256, $3,733,231); Hardin's name is listed in the Award Abstract 9315256, which states, "Mosaic has rapidly grown . . . 300,000 copies of the software have been distributed to date." PI Bruce Schatz, "Building the Interspace: Digital Library Infrastructure for a University Engineering Community" (award 9411318, $4,674,232).

18. See the archived document "Common Gateway Interface" (1995) authored by "cgi@ncsa.uiuc.edu" at www.apacheadmin.com/CGI/intro.html. NCSA once maintained extensive documentation on CGI at hoohoo.ncsa.uiuc.edu and now mirrored at apacheadmin.com/CGI/. The date 1995 is inferred from www.w3.org/CGI/, which states that CGI version 1.1 (cited supra) was an update by David Robinson on 16 Oct. 1995.

19. Frederic J. Wendling Oral History, conducted by Thomas J. Misa and Jeffrey R. Yost, NSF (19–20 Feb. 2008): 47, 48.

20. "There was concern that not all institutions would have access to the Internet—because they wanted to make sure that everyone had a fair play at getting their proposal

statuses. So there was the 'big access to the Internet' conundrum, and making sure that we didn't put something out there that was make an unlevel playing field if you didn't have access to the Internet." Carolyn L. Miller Oral History, conducted by Jeffrey R. Yost, Hyattsville, MD (21 Mar. 2012): 22–23.

21. "Neal [Lane] showed his understanding of the federal bureaucracy and what dysfunctions it's sometimes prey to . . . He said, yes, I want you to try it. I'll give you more money to work on it. And it's all right if you fail," as George Strawn (later NSF CIO) confirms Wendling's account. Strawn Oral History, 7.

22. Lynn Conway, "The Computer Design Process: A Proposed Plan for ACS," *IBM Advanced Computing Systems* (6 Aug. 1968) at tinyurl.com/noam7dj.

23. Elliot I. Organick, *The Multics System: An Examination of Its Structure* (Cambridge, MA: MIT Press, 1972). For computer-security experts David Bell and Len LaPadula, Organick's book introduced them to operating system theory. In subsequent years, David Bell's first request when starting a new project was "Where's the Organick on this operating system?" See David Elliott Bell Oral History, conducted by Jeffrey R. Yost, Reston, VA (24 Sept. 2012): 34, Charles Babbage Institute OH411 at purl.umn.edu/144024.

24. Thomas Weber (chair) et al., "NSF Task Force on Electronic Proposal Processing" (n.d.). Weber's CV is at tinyurl.com/o2pxxap. Muhlbauer was listed in the *1995–96 U.S. Government Manual* (Washington, DC: GPO, 1995), 639–45, as NSF's deputy chief financial officer and director of the Division of Financial Management. Constance McLindon Oral History, conducted by Jeffrey R. Yost, Reston, VA (22 Sept. 2009): 9; Connie McLindon, phone interview with Thomas J. Misa (4 Dec. 2006), confirms her explicit attention to internal NSF "buy-in." Wendling emphasized that "the selling with Neal Lane occurred before the Weber document [and] not as a result of the Weber document . . . Neal approved the project but wanted buy-in both internally and externally, and that's why Connie had Tom [Weber] do that document." Wendling Oral History, 113.

25. Miller Oral History, 29.

26. Wendling to McLindon, 8 Feb. 1994 and 15 Feb. 1994 (emails) in office of NSF historian.

27. Wendling Oral History, 113–15; "Do you mind if we call it FastLane, after you?" is McLindon's recollection. McLindon Oral History, 9.

28. FastLane is also a software engineering algorithm described in Jons-Tobias Wamhoff et al., "FastLane: Improving Performance of Software Transactional Memory for Low Thread Counts," *SIGPLAN Notices* 48, no. 8 (Feb. 2013): 113–22.

29. Wendling Oral History, 57; Miller Oral History, 27.

30. Several agency insiders attributed authorship of this document to Fred Wendling, but he insists that Compuware drafted the document with input from him.

31. Wendling Oral History, 54, 56.

32. Craig Robinson Oral History, conducted by Jeffrey R. Yost, NSF (2 Apr. 2009): 8–11.

33. The components left in outline form in April 1994 were requests for no-cost extensions, cash transaction requests, quarterly federal cash transaction reports, and an information access and query system.

34. Wendling Oral History, 54.

35. Seybold, *Customers.com*, quote p. 164.

36. "NSF began to explore IT research partnerships with Federal mission agencies in 1994, when 12 agencies jointly contributed to development of the first popular web browser, Mosaic," said Larry Brandt, NSF program manager. Quoted in "NSF Creates Partnership to Further Digital Government," NSF press release 00-31 (15 May 2000) at www.nsf.gov

/od/lpa/news/press/00/pr0031.htm, and Melvyn Ciment, "A Personal History of the NSF Digital Government Program," *Communications of the ACM* 46, no. 1 (Jan. 2003): 69–70.

37. Wendling Oral History, 63.

38. Berkeley's DBM was a variant of UNIX guru Ken Thompson's DataBase Manager, whereas SQL (Structured Query Language) implemented Edgar Codd's relational database model.

39. Brian Slator Oral History, conducted by Thomas J. Misa and Jeffrey R. Yost, North Dakota State University (29 Apr. 2008): 5; Tom Cheesbrough Oral History, conducted by Thomas J. Misa, South Dakota State University (30 Apr. 2008): 4.

40. Robert Spector, *Amazon.com: Get Big Fast*, rev. ed. (New York: HarperBusiness, 2002), 47–60, quote p. 50. There is some uncertainty about Amazon's first public book sale: the company's PR site at tinyurl.com/60jwdq lists Douglas R. Hofstadter's *Fluid Concepts and Creative Analogies* (1995) as the first book sold in July 1995, but the purchaser, John Wainwright, states the event was in April 1995 at tinyurl.com/njxuxsk. See also Megan Garber, "Here Is the First Book Ever Ordered on Amazon," *Atlantic* (31 Oct. 2012) at tinyurl.com/d7qmzpu.

41. Rich Schneider Oral History, conducted by Jeffrey R. Yost, NSF (8–9 Mar. 2010): quote p. 7.

42. Miller Oral History, 28.

43. Schneider Oral History, 8.

44. Schneider Oral History, 10.

45. Schneider Oral History, 12; Wendling Oral History, 59.

46. See Ernesto Nebel and Larry Masinter, "Form-based File Upload in HTML," Network Working Group Request for Comments 1867 (Nov. 1995) at www.ietf.org/rfc/rfc1867.txt. Schneider Oral History, 11–12; Wendling Oral History, 59.

47. University of Washington FastLane FAQ (n.d.) at tinyurl.com/k2k2k8e. Northwestern University concurred: "Currently, Netscape 3.0 supports all of FastLane's basic features, but other vendors may provide a qualified browser in the future" at www.research.northwestern.edu/osr/fastlane.html (accessed May 2013; no longer available).

48. See Jason Hunter and William Crawford, *Java Servlet Programming*, 2nd ed. (Sebastopol CA: O'Reilly Media, 2001), 117, and the UCLA PowerPoint presentation "FastLane for Beginners: Research" (2000) at www.research.ucla.edu/slides/Fast101FS.ppt.

49. Miller Oral History, 43.

50. Beverly Sherman Oral History, conducted by Jeffrey R. Yost, NSF (2 Apr. 2009): 29, 35.

51. See Henrik Schmiediche, "Creating a Free PDFWriter Using Ghostscript" (2004) at tinyurl.com/ypaunj.

52. Miller Oral History, 43.

53. In 1999 Congress passed the Federal Financial Assistance Management Improvement Act (Public Law 106-107), which mandated streamlining of federal grant making, while the Government Paperwork Elimination Act (Public Law 105-277) required many forms of electronic government transactions by 2003.

54. See U.S. Government Accountability Office, "Grants Management: Grants.gov Has Systematic Weaknesses That Require Attention," GAO-09-589 (15 July 2009) at www.gao.gov/assets/300/292335.pdf; "Grants Management: Additional Actions Needed to Streamline and Simplify Processes," GAO-05-335 (18 Apr. 18 2005) at www.gao.gov/assets/250/246040.pdf; and "Grants Management: Grantees' Concerns with Efforts to Streamline and Simplify Processes," GAO-06-566 (28 July 2006) at www.gao.gov/assets/260/250911.pdf.

55. Wendling Oral History, 82.

56. On IBM's midrange computers developed in Rochester, MN, see Thomas J. Misa, *Digital State: The Story of Minnesota's Computing Industry* (Minneapolis: University of Minnesota Press, 2013), chapter 6.

57. Brad Linder, "Google Chrome Adds Support for Hotmail" (29 Jan. 2009) at downloadsquad.switched.com/2009/01/29/google-chrome-adds-support-for-hotmail/ (accessed June 2013; no longer available); Chromium Code Reviews, "Issue 19025: Add Support for UA Spoofing, and Spoof Safari's UA String When Loading URLs" at codereview.chromium.org/19025.

58. Robinson Oral History, 10–14.

59. Neal Lane, "All Hands Meeting" speech (Jan. 1996) in NSF historian's office; Miller Oral History, 32–33.

60. Miller Oral History, 32–33.

61. See Paul Hill, *EPSCoR 2030: A Report to the National Science Foundation* (Arlington, VA: NSF, 2012).

62. Miller Oral History, 19, 21.

63. Georgette Sakumoto Oral History, conducted by Jeffrey R. Yost, University of Hawaii at Manoa (15 Nov. 2011): 11.

64. Miller Oral History, 43, 44. We interviewed two team members. For Baisey-Thomas as a FastLane and later Grants.gov trainer, see her bio at tinyurl.com/n8pvhht.

65. Some of our interviews recalled that PDF Writer did the trick, but a 2001 document indicates otherwise (with warnings for both Windows and Mac users): "Adobe Reader cannot produce PDF files. You must use the Adobe Distiller component of Adobe Acrobat. Note: Do Not Use PDFWriter!" See "Creating FastLane PDF Files" at www.nsf.gov/pubs/policydocs/grantsgovguide0113.pdf and at www.fastlane.nsf.gov/documents/pdf_create/pdfcreate_05a.jsp. Misa 10 Apr. 2000 notes archived in CBI Project Files—FastLane.

66. Schneider Oral History, 19–20.

67. Miller Oral History, 25.

68. Wendling Oral History, 119.

69. Miller Oral History, 46.

70. Andrea T. Norris Oral History, conducted by Jeffrey R. Yost, NSF (25 July 2011): 13.

71. Elizabeth VanderPutten Oral History, conducted by Jeffrey R. Yost, NSF (23 Feb. 2011): 5; Robinson Oral History, 13.

72. "FastLane: Working toward a Paperless Proposal Submission and Review Process," *NSF Division of Ocean Sciences Newsletter* (Fall 1999) at www.nsf.gov/pubs/2000/nsf0012/start.htm.

73. Robinson Oral History, 12.

74. The intermediate usage figures are 4% (1997), 15% (1998), and 40% (1999).

75. Jean Feldman Oral History, conducted by Jeffrey R. Yost, NSF (22 Apr. 2011): 18.

76. Robinson Oral History, 4.

77. "I had been on FastLane for five years there, and I was just really burnt-out. I had been working more than 50-, 60-hour weeks, and traveling, and I was having twins, and I just knew that there's no way I could keep with FastLane and keep up my home life." Miller Oral History, 71.

78. Robinson Oral History, 26.

79. Schmitz in NSF directory (1996) at www.nsf.gov/pubs/stis1996/phnorg/phnorg.txt.

80. Robinson Oral History, 27.

81. Robinson Oral History, 6, 11, 14.

82. Robinson Oral History, 18–19.

83. Robinson Oral History, 20.

84. "What's New in FastLane: Ten Year Anniversary of First Transaction Submitted via FastLane," NSF press release (Mar. 2005) from NSF FastLane News Archive (CBI Project Files—FastLane).

85. Robinson Oral History, 28–29.

86. James Perkins Oral History, conducted by Jeffrey R. Yost, Jackson State University (15 July 2008): 7.

87. LaVerne D. Hess Oral History, conducted by Jeffrey R. Yost, NSF (10 Mar. 2010): 11.

CHAPTER FOUR: **Principal Investigators as Lead Users**

1. The National Science Foundation, *Proposal and Award and Procedures Guide*, NSF 07-140 (Apr. 2007): 2. Financial challenges at many U.S. universities have led to an even greater focus on sponsored research to help address budgetary shortfalls. This has likely resulted in more PIs and more proposal submissions per PI.

2. Current and past NSF-funded projects at www.nsf.gov/awardsearch/.

3. The vast majority of the NSF PI community is at U.S. universities and colleges, where we targeted our interviews. We gained some information on the small percentage of PIs outside universities and colleges through our interviews with NSF program officers, including those leading programs to fund small businesses (see, e.g., Joe Hennessey and Bruce Hamilton interviews in the dataset archived at www.cbi.umn.edu/oh/fastlane).

4. Gregory Miller Oral History, conducted by Jeffrey R. Yost, University of Washington (5 Oct. 2010): 4.

5. These offices go by many different names (and acronyms) at universities (such as Sponsored Research Administration, Sponsored Programs Office, Office of Sponsored Research, Office of Research Services, and Office of Research Development). Some schools' programs have changed name over time. For clarity, except within quotations, they are referred to uniformly in this book as *sponsored research offices*, or SROs.

6. Arthur Glenberg Oral History, conducted by Jeffrey R. Yost, Arizona State University (9 Mar. 2011): 3; Edward Wink Oral History, conducted by Jonathan Clemens, University of Minnesota (25 Apr. 2012): 6. Two additional interviewees, who requested confidentiality, also cited use of ditto or mimeograph machines.

7. Among more than a dozen instances, see Charles Weatherford Oral History, conducted by Jeffrey R. Yost, Florida A&M University (28 Nov. 2011): 4; C. Suzanne Iacono Oral History, conducted by Jeffrey R. Yost, NSF (11 Mar. 2010): 4; Tom Cheesbrough Oral History, conducted by Thomas J. Misa, South Dakota State University (30 Apr. 2008): 5; Emi Ito Oral History, conducted by Jonathan S. Clemens, University of Minnesota (7 May 2012): 3; Penny Kukuk Oral History, conducted by Jeffrey R. Yost, University of Montana (18 Oct. 2011): 4; and John Taylor, University of California, Berkeley, online interview collected between July 2011 and June 2012.

8. Eric Hellstrom Oral History, conducted by Jeffrey R. Yost, Florida State University (2 Dec. 2011): 5; Brad R. Weiner Oral History, conducted by Thomas J. Misa, University of Puerto Rico (17 Nov. 2011): 17.

9. "At Yale, they had somebody who would actually fly there and various departments could deliver their packages to this person, and then that person would take them all on a plane to Washington," according to Scott P. Robertson Oral History, conducted by Jeffrey R. Yost, University of Hawaii at Manoa (16 Nov. 2011): 4–5. One PI at the highly respected University of Washington noted, "When I was at Yale, program officers would, in fact,

come to you. I'm not sure if I've seen a program officer in a long time." Kristiina Vogt Oral History, conducted by Jeffrey R. Yost, University of Washington (6 Oct. 2010): 17.

10. Steven N. Robinow Oral History, conducted by Jeffrey R. Yost, University of Hawaii at Manoa (16 Nov. 2011): 4; Clifford W. Morden Oral History, conducted by Jeffrey R. Yost, University of Hawaii at Manoa (18 Nov. 2011): 9; Georgette Sakumoto Oral History, conducted by Jeffrey R. Yost, University of Hawaii at Manoa (15 Nov. 2011): 10.

11. Michael Castellini Oral History, conducted by Thomas J. Misa, University of Alaska Fairbanks (29 Sept. 2011): 6.

12. Daryl Domning Oral History, conducted by Jeffrey R. Yost, Howard University (26 Jan. 2012): 4; Veronica Thomas Oral History, conducted by Jeffrey R. Yost, Howard University (26 Jan. 2012): 4.

13. Dominic P. Clemence Oral History, conducted by Jeffrey R. Yost, North Carolina A&T State University (30 Jan. 2009): 5.

14. Brian Slator Oral History, conducted by Thomas J. Misa and Jeffrey R. Yost, North Dakota State University (29 Apr. 2008): 5; Cheesbrough Oral History, 4.

15. Craig Lent Oral History, conducted by Jeffrey R. Yost, University of Notre Dame (15 Feb. 2012): 3.

16. Dogan Comez Oral History, conducted by Thomas J. Misa, North Dakota State University (28 Apr. 2008): 8.

17. FY1998 ran from 1 July 1997 to 30 June 1998. NSF's fiscal year begins on 1 July of the prior calendar year.

18. Cheesbrough Oral History, 6.

19. "FastLane: Working toward a Paperless Proposal Submission and Review Process," *NSF Division of Ocean Sciences Newsletter* (Fall 1999) at www.nsf.gov/pubs/2000/nsf0012/nsf0012.pdf.

20. Michael Maroney Oral History, conducted by Jeffrey R. Yost, University of Massachusetts Amherst (13 Sept. 2011): 6.

21. William D. Carlson Oral History, conducted by Jeffrey R. Yost, University of Texas at Austin (1 Apr. 2010): 4.

22. James Perkins Oral History, conducted by Jeffrey R. Yost, Jackson State University (15 July 2008): 4.

23. Sandra Romano Oral History, conducted by Thomas J. Misa, University of the Virgin Islands (21 Nov. 2011): 30.

24. Neville Kallenbach Oral History, conducted by Jeffrey R. Yost, New York University (6 Mar. 2012): 5–6; Alexander Lappin Oral History, conducted by Jeffrey R. Yost, University of Notre Dame (16 Feb. 2012): 6.

25. Kallenbach Oral History, 12.

26. NSF, *Accountability Report FY2000* (2001): 14.

27. To the extent any qualitative data on early FastLane users was collected, it either did not survive or, despite significant effort, we could not gain access to it.

28. Anna Prentiss Oral History, conducted by Jeffrey R. Yost, University of Montana (17 Oct. 2011): 3.

29. Mary Berry Oral History, conducted by Thomas J. Misa, University of South Dakota (1 May 2008): 11.

30. Raymond Bernor Oral History, conducted by Jeffrey R. Yost, Howard University (25 Jan. 2012): 6. Numerous critical comments about FastLane's hierarchical structure include those in Keith Bowman Oral History, conducted by Jeffrey R. Yost, Purdue University (28 July 2009).

31. Richard Collins Oral History, conducted by Thomas J. Misa, University of Alaska Fairbanks (27 Sept. 2011): 25.

32. William Stockwell Oral History, conducted by Jeffrey R. Yost, Howard University (25 Jan. 2012): 8.

33. Philip Reid, University of Washington, online interview collected between July 2011 and June 2012, quote p. 2; Frederic J. Wendling Oral History, conducted by Thomas J. Misa and Jeffrey R. Yost, NSF (19–20 Feb. 2008): 65.

34. See UCLA's "FastLane Proposal Preparation 101" (June 2000), slides 35–44, at www.research.ucla.edu/ocga/sr2/fastln.htm.

35. Sat Narain Gupta Oral History, conducted by Jeffrey R. Yost, University of North Carolina at Greensboro (18 July 2008): 6.

36. G. Miller Oral History, 6.

37. Sherry Hsi, University of California, Berkeley, online interview collected between July 2011 and June 2012, quote p. 3; Weatherford Oral History, 9; Washington Mio Oral History, conducted by Jeffrey R. Yost, Florida State University (2 Dec. 2011): 7; John Kelly Oral History, conducted by Jeffrey R. Yost, North Carolina A&T State University (17 July 2008): 7.

38. Kathryn Moler, Stanford University, online interview collected between July 2011 and June 2012, quote p. 3.

39. Jane Muirhead Oral History, conducted by Thomas J. Misa, University of California, Berkeley (9 Feb. 2009): 22.

40. Jonathan Stebbins Oral History, conducted by Jeffrey R. Yost, Stanford University (7 Dec. 2009): 7.

41. Robinow Oral History, 14.

42. Two interviewees requested anonymity; see also Teresa Robinette Oral History, conducted by Jeffrey R. Yost, Arizona State University (8 Mar. 2011): 9.

43. Martha E. Crosby Oral History, conducted by Jeffrey R. Yost, University of Hawaii at Manoa (14 Nov. 2011): 11.

44. Christopher A. Sneden Oral History, conducted by Jeffrey R. Yost, University of Texas at Austin (31 Mar. 2010): 4.

45. Interviewee requested anonymity.

46. George Bodner, Purdue University, online interview collected between July 2011 and June 2012, quote p. 2.

47. Cathryn Carson Oral History, conducted by Thomas J. Misa and Jeffrey R. Yost, University of California, Berkeley (10 Feb. 2009): 7.

48. Braaten in Norm Braaten and John Ruffolo Oral History, conducted by Jeffrey R. Yost, South Dakota State University (30 Apr. 2008): 20.

49. Camille A. McKayle Oral History, conducted by Thomas J. Misa, University of the Virgin Islands (18 Nov. 2011): 8; Bowman Oral History, 4. One of us (Misa) spent *hours* with non-Adobe options before purchasing Acrobat.

50. Stanley May Oral History, conducted by Jeffrey R. Yost, University of South Dakota (1 May 2008): 5; Neil Maxwell Oral History, conducted by Thomas J. Misa, University of California, Berkeley (9 Feb. 2009): 15.

51. William Clyde Oral History, conducted by Thomas J. Misa, University of New Hampshire (14 Sept. 2011): 7.

52. Craig Robinson Oral History, conducted by Jeffrey R. Yost, NSF (2 Apr. 2009): 18–19.

53. Hellstrom Oral History, 8. Hellstrom was at Wisconsin–Madison from 1985 to 2006.

54. Gregory E. Ravizza Oral History, conducted by Jeffrey R. Yost, University of Hawaii at Manoa (16 Nov. 2011): 8.

55. David Fitch Oral History, conducted by Jeffrey R. Yost, New York University (8 Mar. 2012): 6–7.

56. Joseph P. Hayes Oral History, conducted by Jeffrey R. Yost, New York University (7 Mar. 2012): 7.

57. Jesse V. Johnson Oral History, conducted by Thomas J. Misa, University of Montana (18 Oct. 2011): 13.

58. See the NSF report *Women, Minorities, and Persons with Disabilities in Science and Engineering*, NSF 13-304 (Feb. 2013), especially "Black doctorate recipients with HBCU baccalaureate origins" at tinyurl.com/qcjlx53.

59. See the HBCU-UP program announcements, with anticipated funding levels, at www.nsf.gov/pubs/2009/nsf09512/nsf09512.htm and www.nsf.gov/pubs/2011/nsf11519/nsf11519.htm.

60. Paul Basken, "NSF Seeks New Approach to Helping Minority Students in Science," *Chronicle of Higher Education* (10 Mar. 2010) at tinyurl.com/yan9h4j. FY2013 program solicitations for the *independent* programs, HBCU-UP at www.nsf.gov/pubs/2013/nsf13516/nsf13516.htm; LSAMP at www.nsf.gov/pubs/2012/nsf12564/nsf12564.htm, funded at $20 million; Tribal Colleges at www.nsf.gov/pubs/2013/nsf13572/nsf13572.htm, funded at $6 million; and HBCU-RISE at www.nsf.gov/pubs/2013/nsf13533/nsf13533.htm, funded at $2 million. FY 2014–15 HBCU-UP figures from www.nsf.gov/pubs/2014/nsf14513/nsf14513.htm.

61. Other than these four schools (and the University of the Virgin Islands, a historically black university outside the contiguous United States), other HBCUs had too few NSF PIs with both pre-FastLane and FastLane experience (a requirement for our interviews) to make a site visit economically feasible.

62. Information gathered from NSF Awards Database at www.nsf.gov/awardsearch/ (accessed 5 Apr. 2013) and from visits to five HBCUs.

63. Romano Oral History, 4.

64. Clemence Oral History, 4.

65. Clemence Oral History, 6.

66. Clemence Oral History, 6.

67. Our use of the NSF database might have skewed our assessment of the level and difficulties of IT infrastructure at HBCU and other schools; we had access only to successful NSF PIs (unsuccessful proposals are *not* made public). We could not readily identify researchers who submitted to NSF but were never funded or those who may have chosen not to submit to NSF because of insufficient IT resources or support.

68. James D. Savage, *Funding Science in America: Congress, Universities, and the Politics of the Academic Pork Barrel* (Cambridge: Cambridge University Press, 1999): 61.

69. Christine M. Matthews, *U.S. National Science Foundation: Experimental Program to Stimulate Competitive Research (EPSCoR)* (15 Jan. 2009).

70. Matthews, *U.S. National Science Foundation*; Paul Hill, *EPSCoR 2030: A Report to the National Science Foundation* (Apr. 2012): 1–4. For a 2013 map indicating when individual states or territories joined EPSCoR, see www.nsf.gov/od/iia/programs/epscor/images/cohortmap2013.pdf.

71. Hill, *EPSCoR 2030*, 4.

72. Hill, *EPSCoR 2030*, 4–10.

73. Braaten and Ruffolo Oral History, 7.

74. Jeffrey Clark Oral History, conducted by Jeffrey R. Yost, North Dakota State Uni-

versity (28 Apr. 2008): 14–15. Clark recalled: "I called the reviews came in, it was competitive, the reviews I thought were pretty good, I was kind of surprised it didn't get funded. I called and talked to the program director and said even with the reviews, I thought we'd at least have gotten EPSCoR consideration, which could be enough to give you funding. There was hesitation at the other end and he said, 'Ahhhh, yeah. I'll have to look into that.' He forgot. He didn't get many EPSCoR proposals, and he said I'll get back to you in a couple weeks. We didn't get funded."

75. Peter Schweitzer Oral History, conducted by Thomas J. Misa, University of Alaska Fairbanks (27 Sept. 2011): 20.

76. J. B. Alexander Ross Oral History, conducted by Thomas J. Misa, University of Montana (18 Oct. 2011): 20–21. And additional interviews at EPSCoR schools where PIs requested anonymity.

77. Robinson Oral History, 17. Robinson stated that NSF visited many tribal colleges for FastLane presentations and workshops, as well as other minority-serving institutions (including HBCUs).

78. Raymond Ford Oral History, conducted by Jeffrey R. Yost, University of Montana (17 Oct. 2011): 22; Weiner Oral History, 7. Several EPSCoR schools also faced challenges with unreliable or slow networking at the launch of FastLane, including University of Puerto Rico.

79. Ford Oral History, 22.

80. Several NSF PIs who collaborate with tribal colleges noted they often lacked resources to support sponsored research.

81. Melody Kadenko Oral History, conducted by Jeffrey R. Yost, University of Washington (7 Oct. 2010): 9.

82. Kadenko Oral History, 8–10.

83. Claire Christopherson Oral History, conducted by Thomas J. Misa, University of Massachusetts Amherst (12 Sept. 2011): 5–6.

84. Hayes Oral History, 4–6.

85. Joseph E. Hennessey Oral History, conducted by Jeffrey R. Yost, NSF (22 Feb. 2011): 7.

86. Hennessey Oral History, 7.

87. Alan Kotok, "Is Grants.gov in the FastLane?" *AAAS Science Careers* (9 Mar. 2007) at tinyurl.com/q5aqey4.

88. Stacia Sower Oral History, conducted by Thomas J. Misa, University of New Hampshire (14 Sept. 2011): 16; Maxwell Oral History, 12; Marilyn Brandt Oral History, conducted by Thomas J. Misa, University of the Virgin Islands (21 Nov. 2011): 9.

89. Beverly Sherman Oral History, conducted by Jeffrey R. Yost, NSF (2 Apr. 2009): 19; Carolyn L. Miller Oral History, conducted by Jeffrey R. Yost, Hyattsville, MD (21 Mar. 2012): 44.

90. C. Miller Oral History, 20.

91. Kadenko Oral History, 14.

92. Herbert Strauss, Stanford University, online interview collected between July 2011 and June 2012, p. 2.

93. Nicholas Cogan Oral History, conducted by Jeffrey R. Yost, Florida State University (1 Dec. 2011): 7; report on piloting annual reports published in *Science* 280, no. 5366 (15 May 1998): 975.

94. McKayle Oral History, 14–15.

95. Bowman Oral History, 13.

96. Brandt Oral History, 14.

97. Dawn Meredith Oral History, conducted by Thomas J. Misa, University of New Hampshire (15 Sept. 2011): 13.

98. Greg Newby Oral History, conducted by Thomas J. Misa, University of Alaska Fairbanks (28 Sept. 2011): 28.

99. Interviewee at an HBCU requested anonymity.

100. Matthew Ohland Oral History, conducted by Jeffrey R. Yost, Purdue University (28 July 2009): 21.

CHAPTER FIVE: Research Administrators as Lead Users

1. For consistency and clarity, SRO is used as a general abbreviation for university central sponsored research offices throughout the text (except, at times, for quoted material).

2. The term "PI" or principal investigator is used in this chapter to refer to both current PIs and those researchers who apply to be PIs on sponsored projects. (Many have been PIs on past projects.) All our interviews with researchers (generally faculty) at universities on this project were with past or present NSF PIs, since we identified interview candidates from the NSF awards database. NSF provides no information on unfunded proposal applicants. Officially submitting refers to the institutional *authority* to submit (not to those who physically mailed proposals in the paper-based days).

3. Pamela A. Webb Oral History, conducted by Jonathan Clemens, University of Minnesota (27 Feb. 2012): 4. Webb, at UCLA at the time, recalled, "I'm pretty sure that I submitted the first live FastLane application. Neil Maxwell at Berkeley . . . submitted the first dead data application from FastLane." Maxwell claimed that his application was the first. Neil Maxwell Oral History, conducted by Thomas J. Misa, University of California, Berkeley (9 Feb. 2009), unrecorded comments.

4. Unless otherwise identified as department-level grant administrators, all discussion of "grant administrators" in this chapter refers to central SRO grant administrators, not those located within academic departments.

5. Courtney Frazier Swaney Oral History, conducted by Jeffrey R. Yost, University of Texas (30 Mar. 2010): 3; Matt Hawkins Oral History, conducted by Jeffrey R. Yost, University of Washington (4 Oct. 2010): 3; Elena V. Mota Oral History, conducted by Jeffrey R. Yost, University of Texas at Austin (30 Mar. 2010): 3; Reggie W. Crim Oral History, conducted by Jeffrey R. Yost, University of Texas at Austin (30 Mar. 2010): 3; Susan Mundt Oral History, conducted by Jeffrey R. Yost, University of Arizona (15 Mar. 2011): 3; Pete Lester Oral History, conducted by Jeffrey R. Yost, University of New Hampshire (14 Sept. 2011): 3; Julie Nash Oral History, conducted by Jeffrey R. Yost, North Dakota State University (28 Apr. 2008): 3; Greg Simpson Oral History, conducted by Thomas J. Misa, University of Alaska Fairbanks (29 Sept. 2011): 3.

6. Carol Sprague Oral History, conducted by Jeffrey R. Yost, University of Massachusetts Amherst (13 Sept. 2011): 4–5.

7. Linda Campbell Oral History, conducted by Jeffrey R. Yost, Santa Clara University (13 Aug. 2008): 4.

8. Maxwell Oral History, 5–6.

9. Vicki Krell Oral History, conducted by Jeffrey R. Yost, Arizona State University (7 Mar. 2011): 9.

10. See National Academy of Sciences, "About FDP" (n.d.) at tinyurl.com/khr709e.

11. Maxwell Oral History, 8–9.

12. Braaten in Norm Braaten and John Ruffolo Oral History, conducted by Jeffrey R. Yost, South Dakota State University (30 Apr. 2008): 24–25.

13. Charna Howson Oral History, conducted by Jeffrey R. Yost, University of North Carolina at Greensboro (18 July 2008): 6, 7.

14. Tiffany Blackman, Harvard University, online interview collected between July 2011 and June 2012, quote p. 3.

15. Six interviewees, including one NSF staff, recalled the epithet "SlowLane." See, e.g., Andrew Gray Oral History, conducted by Thomas J. Misa, University of Alaska Fairbanks (27 Sept. 2011): 13; Jess Zimmerman Oral History, conducted by Thomas J. Misa, University of Puerto Rico (17 Nov. 2011): 10; Michael Overton Oral History, conducted by Jeffrey R. Yost, New York University (6 Mar. 2012): 4; Claudia Neuhauser Oral History, conducted by Jonathan Clemens, University of Minnesota (16 May 2012): 4.

16. Gray Oral History, 13–14.

17. Jonathan Yeargan Oral History, conducted by Jeffrey R. Yost, Florida State University (30 Nov. 2011): 5; David Wilson Oral History, conducted by Jeffrey R. Yost, Jackson State University (14 July 2008): 8; Naomi Mitake Oral History, conducted by Jeffrey R. Yost, University of Hawaii at Manoa (15 Nov. 2011): 8.

18. Hawkins Oral History, 4.

19. Krell Oral History, 7.

20. Federal granting agencies require grantee institutions to draw funds in relatively small increments for budgetary and cash flow reasons.

21. Mona Weer Oral History, conducted by Jeffrey R. Yost, University of Montana (17 Oct. 2011): 4.

22. See FAQ on ASAP at www.fms.treas.gov/asap/background.html; *NSF Grant Policy Manual*, NSF 05-131 (July 2005), chapter 4 at tinyurl.com/pdd5hcy.

23. Mary Jo Hershly Oral History, conducted by Jeffrey R. Yost, University of Washington (7 Oct. 2010): 4.

24. Hershly Oral History, 4.

25. Howard University has a major medical school, which substantially boosts its sponsored research profile—particularly to NIH but also, through allied scientific fields, to NSF.

26. Rita Presley Oral History, conducted by Jeffrey R. Yost, Jackson State University (14 July 2008): 4.

27. Wilson Oral History, 7; Beverly Sherman Oral History, conducted by Jeffrey R. Yost, NSF (2 Apr. 2009): 52.

28. This had mixed results. There seems to have been greater cooperation with HBCUs and EPSCoR schools in this regard—perhaps because they recognized the special efforts NSF was making and thus followed NSF's inclusiveness. Local rivalries led some schools not to invite neighboring schools. NSF's Beverly Sherman recalled one month she went to a different school each week for a month and none of them invited other schools. Sherman Oral History, 21.

29. The Mississippi Research Consortium was launched in 1986.

30. Presley Oral History, 11.

31. Presley Oral History, 4.

32. Wilson Oral History, 20, 5, 7.

33. Wilson Oral History, 7–8.

34. Wilson Oral History, 12. NSF staff would also put on workshops at Jackson State targeting faculty.

35. Presley Oral History, 11–12.

36. Between 1990 and 2001, Jackson State averaged 2.4 new NSF awards per year; between 2002 and 2014, 6.2 per year. See www.nsf.gov/awardsearch/.

37. In fact, aside from Jackson State, SRO staff at other HBCUs restricted access for their interviews. Overall, more than three-quarters of SRO managers and staff at HBCUs selected restricted access options (preventing quotes, attribution of names, or schools).

38. Howson Oral History, 7.

39. Two of the five HBCUs were also EPSCoR schools—Jackson State University and University of the Virgin Islands.

40. Braaten and Ruffolo Oral History, 7.

41. Braaten and Ruffolo Oral History, 14.

42. Braaten and Ruffolo Oral History, 5. In 2007 Braaten moved from the University of Nebraska SRO to join the South Dakota State University SRO.

43. Braaten and Ruffolo Oral History, 33.

44. Braaten and Ruffolo Oral History, 34. The SRO, of course, still provided the institutional authorization.

45. Interviewees specifically recalled EPSCoR grants improving FastLane institutional readiness and adoption at University of Montana, University of Alaska Fairbanks, University of Puerto Rico, University of the Virgin Islands, and University of Hawaii.

46. See "Best and Worst Run States in America" *24/7 Wall St.* (27 Nov. 2012) at tinyurl .com/mnzl9n2.

47. Interviewees requested confidentiality.

48. Krell Oral History, 15.

49. Krell Oral History, 23.

50. John V. Lombardi, Elizabeth D. Phillips, Craig W. Abbey, and Diane D. Craig, *The Top Research Universities: 2011 Annual Report* (Tempe: Center for Measuring Performance, Arizona State University, 2011) at mup.asu.edu/research2011.pdf. Establishing cause and effect is not possible (at least from our data), though given ASU's rapid growth in sponsored funding it might well have offset its investment in additional research administration staff through indirect cost recovery.

51. Many SRO staff work especially long hours on key agency deadline days.

52. As of 5 Apr. 2013, according to www.ors.hawaii.edu.

53. Yaa-Yin Fong Oral History, conducted by Jeffrey R. Yost, University of Hawaii at Manoa (18 Nov. 2011): 4.

54. The University of Hawaii at Manoa is one of roughly a dozen schools in this research administration software development consortium.

55. Fong Oral History, 4.

56. H. Ronald Riggs Oral History, conducted by Jeffrey R. Yost, University of Hawaii at Manoa (17 Nov. 2011): 11.

57. Gary Barnes Oral History, conducted by Jeffrey R. Yost, University of Hawaii at Manoa (14 Nov. 2011): 20.

58. Past ERA front end would focus on pieces, such as electronic routing and approval, but not system-to-system for agencies. (The many different systems made system-to-system extremely difficult, if not technically or financially impossible.)

59. Some PIs and SRO staff brought up NSPIREs; in general, most like the system and find it far easier and more stable than Grants.gov but not as easy and user friendly as FastLane.

60. University of Hawaii's geography and research specializations appear to result in faculty applying to a greater number of sponsored funding agencies and entities than other schools of similar size.

61. See www.cayuse.com. Cayuse has been developing ERA software for more than a decade.

62. Rich Schneider Oral History, conducted by Jeffrey R. Yost, NSF (8–9 Mar. 2010): 34–35.

63. Marianne Siroker Oral History, conducted by Jeffrey R. Yost, Stanford University (8 Dec. 2009): 10.

64. Terri Coslet Oral History, conducted by Jeffrey R. Yost, University of Montana (17 Oct. 2011): 10.

65. Amanda Hamaker Oral History, conducted by Jeffrey R. Yost, Purdue University (23 July 2009): 15.

66. Julie Wammack Oral History, conducted by Jeffrey R. Yost, Florida State University (30 Nov. 2011): 10.

67. Joseph P. Hayes Oral History, conducted by Jeffrey R. Yost, New York University (7 Mar. 2012): 16.

68. Sinh Simmons Oral History, conducted by Jeffrey R. Yost, University of Washington (5 Oct. 2010): 5–6.

69. Karen Henry Oral History, conducted by Jeffrey R. Yost, Boise State University (20 Oct. 2011): 20.

70. Krell Oral History, 24.

CHAPTER SIX: **NSF Staff as Legacy Users**

1. Andrea T. Norris Oral History, conducted by Jeffrey R. Yost, NSF (25 July 2011): 12.

2. Craig Robinson Oral History, conducted by Jeffrey R. Yost, NSF (2 Apr. 2009): 44.

3. Robinson Oral History, 22, 32. In 2014, NSF employed 1,300 career and temporary employees, or rotators; see Division of Human Resource Management at www.nsf.gov/oirm/hrm/ (as of Jan. 2015).

4. Carolyn L. Miller Oral History, conducted by Jeffrey R. Yost, Hyattsville, MD (21 Mar. 2012): 50.

5. R. Corby Hovis Oral History, conducted by Jeffrey R. Yost, NSF (18 Feb. 2011): 12. eJacket first came into widespread use throughout NSF for proposal declinations in 2004.

6. Norris Oral History, 32. According to Norris, at the time eJacket for declinations was under development, there were 70% declinations and 30% awards. In recent years, funded proposals have varied between 20 and 25%.

7. Hovis Oral History, 13.

8. Christopher W. Stark Oral History, conducted by Jeffrey R. Yost, NSF (12 Mar. 2010): 18.

9. Sonya Mallinoff Oral History, conducted by Jeffrey R. Yost, NSF (17 Feb. 2011): 9–10.

10. Renata Thompson Oral History, conducted by Jeffrey R. Yost, NSF (26 July 2010): 5; Judith Verbeke Oral History, conducted by Jeffrey R. Yost, NSF (8 Mar. 2010): 11; Richard Smith Oral History, conducted by Jeffrey R. Yost, NSF (19 Apr. 2011): 12.

11. Elizabeth Rom Oral History, conducted by Jeffrey R. Yost, NSF (22 Feb. 2011): 3. Program officers emphasized the considerable stress and lost time in searching for proposals.

12. Saran Twombly Oral History, conducted by Jeffrey R. Yost, NSF (19 Apr. 2011): 10.

13. Thomas J. Baerwald Oral History, conducted by Jeffrey R. Yost, NSF (8 Mar. 2010): 7–8; Joanne G. Rodewald Oral History, conducted by Jeffrey R. Yost, NSF (29 July 2011): 19; Miller Oral History, 7. PARS, a system NSF developed in the early-to-mid 1990s, was a reviewer database that also offered form letters for program officers.

14. Baerwald Oral History, 4–5.

15. This was lessened for some midsized panels with NSF's move to Arlington, but hotels continued to be used for many of the larger panels.

16. Miles Boylan Oral History, conducted by Jeffrey R. Yost, NSF (14 Feb. 2011): 8.

17. Boylan Oral History, 8–9.

18. Boylan Oral History, 8.

19. Rom Oral History, 3.

20. Quite possibly DIS contacted and got feedback on the design from a small number of program officers or PIs, but such feedback was not mentioned in any of our interviews.

21. Fae L. Korsmo Oral History, conducted by Jeffrey R. Yost, NSF (10 Mar. 2010): 13.

22. Hovis Oral History, 16–17.

23. Alexandra Isern Oral History, conducted by Jeffrey R. Yost, NSF (18 Feb. 2011): 9.

24. Rebecca Ostertag Oral History, conducted by Jeffrey R. Yost, University of Hawaii at Hilo (21 Nov. 2011): 8; Daniel L. Stein Oral History, conducted by Jeffrey R. Yost, New York University (7 Mar. 2012): 8.

25. Miller Oral History, 50.

26. Jean Feldman Oral History, conducted by Jeffrey R. Yost, NSF (22 Apr. 2011): 24.

27. See the Tribal Colleges and Universities Program at www.nsf.gov/funding/pgm_summ.jsp?pims_id=5483.

28. Feldman Oral History, 26.

29. Isern Oral History, 3–4.

30. Mallinoff Oral History, 14–15.

31. Feldman Oral History, 25.

32. Norris Oral History, 31.

33. Beverly Berger Oral History, conducted by Jeffrey R. Yost, NSF (14 Feb. 2011): 9.

34. Frank P. Scioli Oral History, conducted by Jeffrey R. Yost, NSF (10 Mar. 2010): 9.

35. Norris Oral History, 16.

36. Susan Kemnitzer Oral History, conducted by Jeffrey R. Yost, NSF (15 Feb. 2011): 5.

37. Stark Oral History, 14.

38. Miles Boylan Oral History, conducted by Jeffrey R. Yost, NSF (14 Feb. 2011): 21; Anne-Marie Schmoltner Oral History, conducted by Jeffrey R. Yost, NSF (18 Feb. 2011): 16.

39. Kemnitzer Oral History, 5–6.

40. Eric C. Itsweire Oral History, conducted by Jeffrey R. Yost, NSF (8 Mar. 2010): 9–10.

41. Berger Oral History, 20–21.

42. Some support staff we interviewed had moved to new work roles (which was clearly evident from their promotion in titles); others had remained in lower level positions, and some of these people told us that they had less to do because of eJacket.

43. Conversations we had and the presence of many midcareer women program officers today both indicate improving gender balance at NSF.

44. George Strawn Oral History, conducted by Jeffrey R. Yost, NSF (10 Mar. 2010): 12.

45. Mallinoff Oral History, 3–6.

46. Daphne Marshall Oral History, conducted by Jeffrey R. Yost, NSF (25 July 2011): 11.

47. Thompson Oral History, 8.

48. Marshall Oral History, 7.

49. We interviewed one African American woman support staff member who has been shuffled among basic assignments around the foundation, had seen little advancement, and

has little sense of fulfillment or feeling of contribution with her job. She was highly successful in the paper-based era and found her job fulfilling. (This individual requested that her name be kept confidential and the interview not be public.) Our interviews with program officers and support staff indicate that hers was far from an isolated case. At the very least, some support staff employees did not transition well to computerization in general and to eJacket in particular. We were told that some had left NSF and others had retired—NSF downsized support staff over time by attrition, not by firings or layoffs. Our method of identifying support staff for interviews was asking program officers for the names of staff in their division or other divisions. Though we stressed our desire to speak with support staff with a range of job classifications and assignments, most suggestions we received were people who had succeeded in getting promoted to support-staff managerial posts. More than two-thirds of support staff we invited to participate did not respond or declined to be interviewed (a significantly lower participation rate than program officers). Support staff with job titles that appeared to be lower level were far less likely to agree to be interviewed.

50. Joseph F. Burt Oral History, conducted by Jeffrey R. Yost, NSF (26 July 2011): 20.

51. Burt Oral History, 21.

52. Paul Morris Oral History, conducted by Jeffrey R. Yost, NSF (20 Apr. 2011): 3–10.

53. Morris Oral History, 11–18, quote pp. 17–18.

54. In March 2013 NSF began automated compliance checking on 10 required sections of all proposals, and by January 2015 there were 24 items subject to compliance checking. See "Automated Compliance Checking of NSF Proposals" at www.nsf.gov/bfa/dias/policy/autocompliance.jsp and the 24-item checklist at www.nsf.gov/bfa/dias/policy/autocheck/compliancechecks_jan15.pdf.

55. Strawn Oral History, 23. For a computer-assisted means for forming panels, including dealing with "orphans," see Seth Hettich and Michael J. Pazzani. "Mining for Proposal Reviewers: Lessons Learned at the National Science Foundation," *Proceedings of the Twelfth ACM SIGKDD International Conference on Knowledge Discovery and Data Mining* (New York: ACM Press, 2006): 862–71, esp. 865.

56. Strawn Oral History, 19.

57. Strawn Oral History, 13.

58. Erika Rissi Oral History, conducted by Jeffrey R. Yost, NSF (29 July 2011): 3–7.

59. Karen Evans, testimony to U.S. House Subcommittee on Technology, Information Policy, Intergovernmental Relations and the Census hearing on "Electronic Government: A Progress Report on the Successes and Challenges of Government-Wide Information Technology Solutions," Serial No. 108-195 (24 Mar. 2004) at tinyurl.com/oggqk6a; Rissi Oral History, 7–8.

60. Rissi Oral History, 9.

61. Rissi Oral History, 14.

62. Rissi Oral History, 14.

CHAPTER SEVEN: **Legacies, Lessons, and Prospects**

1. Dan Atkins, interview at NSF with Thomas J. Misa (17 Nov. 2006). On CSCW, see Jonathan Grudin, "Computer-Supported Cooperative Work: History and Focus," *IEEE Computer* 27, no. 5 (1994): 19–26.

2. "NSF Names Daniel Atkins to Head New Office of Cyberinfrastructure," NSF press release 06-025 (8 Feb. 2006) at www.nsf.gov/news/news_summ.jsp?cntn_id=105820; the original Atkins report is available at www.nsf.gov/cise/sci/reports/atkins.pdf, quotes pp. 4–5.

3. See Chris Ferguson, "Whose Vision? Whose Values? On Leading Information Ser-

vices in an Era of Persistent Change," in Karin Wittenborg, Chris Ferguson, and Michael A. Keller, eds., *Reflecting on Leadership* (Washington, DC: Council on Library and Information Resources, 2003) at tinyurl.com/ktah8jf. "We have to break the old culture here," was the remark of one insistent advocate of "reengineering" at MIT, according to Rosalind Williams, "'All That Is Solid Melts into Air': Historians of Technology in the Information Revolution," *Technology and Culture* 41, no. 4 (Oct. 2000): 641–68, quote p. 651.

4. Carolyn L. Miller Oral History, conducted by Jeffrey R. Yost, Hyattsville, MD (21 Mar. 2012): 31.

5. David G. Barnes and Christopher J. Fluke, "Incorporating Interactive Three-Dimensional Graphics in Astronomy Research Papers," *New Astronomy* 13, no. 8 (Nov. 2008): 599–605; David G. Barnes et al., "Embedding and Publishing Interactive, 3-Dimensional, Scientific Figures in Portable Document Format (PDF) Files," *PLoS One* 8, no. 9 (2013): e69446.

6. Jeffrey Mervis, "NSF Moves into FastLane to Manage Flow of Grants," *Science* 267, no. 5195 (13 Jan. 1995): quote p. 166; Anne Peterson, "NSF FastLane Goals," *Science* 267 no. 5198 (3 Feb. 1995): 601–2.

7. Sherry Turkle, ed., *Evocative Objects: Things We Think With* (Cambridge, MA: MIT Press, 2007).

8. According to appendix A of the 2000 *NSF Grant Proposal Guide*, at www.nsf.gov/pubs/2000/nsf002/apx_a.htm, just two programs at NSF required a *single* paper proposal copy (IGERT and GRF) while most directorates required 10, 15, or 20 copies of paper proposals.

9. Peter Schweitzer Oral History, conducted by Thomas J. Misa, University of Alaska Fairbanks (27 Sept. 2011): 20.

10. See NSF, "FY1996 Report on the NSF Merit Review System: Section B. Methods of Proposal Review: Subsection on Peer Review and Merit Review" (1997) at www.nsf.gov/nsb/documents/1997/nsb9713/meritrpt.htm; Marc Rothenberg, "Making Judgments about Grant Proposals: A Brief History of the Merit Review Criteria at the National Science Foundation," *Technology & Innovation* 12, no. 3 (2010): 189–95.

11. Erika Rissi Oral History, conducted by Jeffrey R. Yost, NSF (29 July 2011): 10; Joyce Evans Oral History, conducted by Jeffrey R. Yost, NSF (18 Feb. 2011): 13; Joseph E. Hennessey Oral History, conducted by Jeffrey R. Yost, NSF (22 Feb. 2011): 15.

12. Craig Robinson Oral History, conducted by Jeffrey R. Yost, NSF (2 Apr. 2009): 5, 35; Alexander N. Shor Oral History, conducted by Jeffrey R. Yost, University of Hawaii at Manoa (14 Nov. 2011): 15.

13. Robinson Oral History, 39. Robinson also recalled, "There had been an e-jacket before [2002] and it was based in very old software and, again, it had just been static" (22). Another NSF staffer, in a restricted-access interview, remembered using FoxPro database software.

14. David E. Schindel, "Cost Comparison of Panel Review with and without FastLane," NSF Office of Science and Technology Infrastructure (n.d.), document in possession of NSF historian.

15. Contrast Bruce W. Arden, ed., *What Can Be Automated?: The Computer Science and Engineering Research Study (COSERS)* (Cambridge, MA: MIT Press, 1980), with Matti Tedre, "What Should Be Automated?" *ACM Interactions* 15, no. 5 (Sept. 2008): 47–49.

16. See Erich Bloch, interview with NSF historian Marc Rothenberg (11 Feb. 2008).

17. In our interviews, some people were wary of making generalizations or discussing possibly sensitive or controversial topics, such as job losses. At least as often, however, with the unobtrusive digital recorder we used, we experienced just the opposite: that people were

honest, blunt, and explicit in their comments, viewpoints, and perspectives. The institutional review board (IRB) under which we conducted these interviews instructed us to not, in any way, put undue pressure on a NSF staff member who did not wish to be interviewed, which we fully respected.

18. Eric von Hippel, *Democratizing Innovation* (Cambridge, MA: MIT Press, 2005); Nik Franke, User-Driven Innovation," in Mark Dodgson, David Gann, and Nelson Phillips, ed., *The Oxford Handbook of Innovation Management* (Oxford: Oxford University Press, 2014), chapter 5.

19. Paul Morris Oral History, conducted by Jeffrey R. Yost, NSF (20 Apr. 2011): 18. See chapter 6 for a more detailed discussion.

20. Pamela A. Webb Oral History, conducted by Jonathan S. Clemens, University of Minnesota (27 Feb. 2012): 6.

21. Miller Oral History, 32–33.

22. Saran Twombly Oral History, conducted by Jeffrey R. Yost, NSF (19 Apr. 2011): 4.

23. Webb Oral History, 6.

24. Webb Oral History, 6–8.

25. Webb Oral History, 44.

26. Andrea T. Norris Oral History, conducted by Jeffrey R. Yost, NSF (25 July 2011): 7.

27. Scott Borg Oral History, conducted by Jeffrey R. Yost, NSF (17 Feb. 2011): 8.

28. Miller Oral History, 49.

29. See Batya Friedman and Helen Nissenbaum, "Bias in Computer Systems," *ACM Transactions on Information Systems* 14, no. 3 (1996): 330–47; Mary Flanagan, Daniel C. Howe, and Helen Nissenbaum, "Embodying Values in Technology: Theory and Practice," in M. J. van den Joven and J. Weckert, eds., *Information Technology and Moral Philosophy* (Cambridge: Cambridge University Press, 2008), 322–53; and Philip Brey, "Values in Technology and Disclosive Computer Ethics," in Luciano Floridi, ed., *The Cambridge Handbook of Information and Computer Ethics* (Cambridge: Cambridge University Press, 2010), 41–58.

30. "Information wants to be free because it has become so cheap to distribute, copy, and recombine—too cheap to meter. It wants to be expensive because it can be immeasurably valuable to the recipient," in Stewart Brand, *Media Lab: Inventing the Future at MIT* (New York: Viking, 1987), quote p. 202.

31. Jean Feldman Oral History, conducted by Jeffrey R. Yost, NSF (22 Apr. 2011): 4.

32. Feldman Oral History, 2–5.

33. Robinson Oral History, 46; Norris Oral History, 14, 16, 33; William Zamer Oral History, conducted by Jeffrey R. Yost, NSF (20 Apr. 2011): 20.

34. Miller Oral History, 9, 59; Robinson Oral History, 11, 24.

35. Miller Oral History, 45; Feldman Oral History, 16.

36. Webb Oral History, 25. At the University of Puerto Rico, "The EPSCoR runs from the central administration office because it involves multiple campuses. And that office, the Resource Center for Science and Engineering, would function as its own SRO in a way. So there was a lot of expertise resident in there. We built up some local resident expertise that we could always help people with." Brad R. Weiner Oral History, conducted by Thomas J. Misa, University of Puerto Rico (17 Nov. 2011): 13.

37. See Narain Gupta Oral History, conducted by Jeffrey R. Yost, University of North Carolina at Greensboro (18 July 2008): 14, and Alexandra Isern Oral History, conducted by Jeffrey R. Yost, NSF (18 Feb. 2011): 5; history department "guru" identified in an interview requesting anonymity.

38. See http://nspires.nasaprs.com/external.

39. See "Fact Sheet: American Recovery & Reinvestment Act" (26 Mar. 2009) at tinyurl

.com/moph6k. The Grants.gov suspension was announced in "NSF FastLane for New Proposal Submission" (21 May 2009), which stated, "Due to an expected increase in Grants.gov submissions relating to the processing of Recovery Act proposals, the Office of Management and Budget (OMB) has authorized agencies to use alternative methods for proposal submission and acceptance. As you know, NSF is able to accept directly its full complement of proposals, both regular submissions and those additional proposals anticipated under the Recovery Act, using our long-established FastLane capabilities for proposal submission and acceptance. Therefore, in order to assist Grants.gov in the effort to alleviate system strain and increase system capacity, proposers will now be required to prepare and submit proposals to NSF through use of the NSF FastLane system. *Effective immediately, new funding opportunities issued by NSF will exclusively require the use of FastLane to prepare and submit proposals*" [emphasis added] at www.research.uky.edu/arra/nsf.html.

40. See Marc Parry, "Business Software, Built by Colleges for Colleges, Challenges Commercial Giants," *Chronicle of Higher Education* (15 Nov. 2009) at tinyurl.com/ydjnq20.

41. "Free Exchange: The Humble Hero," *Economist* (18 May 2013) at tinyurl.com/ahfnsq8; Paul E. Ceruzzi, "Moore's Law and Technological Determinism: Reflections on the History of Technology," *Technology and Culture* 46 no. 3 (2005) 584–593, quote pp. 588–89.

42. For recent congressional scrutiny, see Jeffrey Mervis, "Battle between NSF and House Science Committee Escalates: How Did It Get This Bad?" *Science* (2 Oct. 2014) at tinyurl.com/knzljmv.

43. Lee Zia Oral History, conducted by Jeffery R. Yost, NSF (15 Feb. 2011), quote p. 5 (huge volume); Neil Maxwell Oral History, conducted by Thomas J. Misa, University of California Berkeley (9 Feb. 2009), quote p. 6 (still here).

44. Thomas Wailgum, "University ERP: Big Mess on Campus," *CIO* (1 May 2005) at tinyurl.com/pdwxk8k; Michael Krigsman, "ERP Train Wrecks, Failures, and Lawsuits," *ZDNet* (19 Jan. 2011) at tinyurl.com/mn4pqyh. Quote from Abbie Lewis, "The Top Three ERP Implementation Disasters," *Sharedserviceslink* (27 Nov. 2013) at tinyurl.com/m6n955g. Even at computer-savvy MIT, a cultural mismatch was evident with the installation of SAP R/3, where "the fundamental problem was that the purchasing software was designed for a corporate social setting . . . not for a university setting," according to Rosalind Williams, "All That Is Solid," 651.

Essay on Sources

As a historical assessment of computing at the National Science Foundation, this book draws on several literatures and bodies of knowledge. There are numerous gaps. To our knowledge, we lack a comparative history of federal grant-making across all agencies as well as a multivolume history of computing inside the disparate branches of the federal government. This book examines one element of this puzzle. Comparison of NSF's computerizing its core grant-making activities with similar processes at the Department of Energy, NASA, National Institutes of Health, or the Department of Defense is simply not feasible until additional research is commissioned and completed. We sorely lack objective and responsible appraisals of the complex, computerized grant-making systems that annually dispense billions of dollars in research funding and consequently shape the future of science, technology, and society. It might seem odd to put it this way, but we have lived through a "digital revolution" but have only the most rudimentary insight into the actual changes it has brought about in our governmental structures, research institutions, and civic practices.

Secondary Sources

The National Science Foundation is understandably a key topic in the history of American science and science policy. Required readings include Daniel J. Kevles, "The National Science Foundation and the Debate over Postwar Research Policy, 1942–1945: A Political Interpretation of 'Science, the Endless Frontier,'" *Isis* 68 (1977): 5–26; Jessica Wang, "Liberals, the Progressive Left, and the Political Economy of Postwar American Science: The National Science Foundation Debate Revisited," *Historical Studies in the Physical and Biological Sciences* 26, no. 1 (1995): 139–66; David M. Hart, *Forged Consensus: Science, Technology, and Economic Policy in the United States, 1921–1953* (Princeton, NJ: Princeton University Press, 1998), 158–64; and Daniel Lee Kleinman, *Politics on the Endless Frontier: Postwar Research Policy in the United States* (Durham, NC: Duke University Press, 1995).

Institutional histories of NSF give essential background and a sense of agency policies. Agency overviews include J. Merton England, *A Patron for Pure Science: The National Science Foundation's Formative Years, 1945–1957* (Washington, DC: NSF, 1982), and George T. Mazuzan, "The National Science Foundation: A Brief History," NSF 88-16 (July 1994). Studies of specific fields include Toby A. Appel, *Shaping Biology: The National Science Foundation and American Biological Research, 1945–1975* (Baltimore: Johns Hopkins University Press, 2000); Otto N. Larsen, *Milestones and Millstones: Social Science at the National Science Foundation, 1945–1991* (New Brunswick, NJ: Transaction, 1992); and Dian Olson Belanger,

Enabling American Innovation: Engineering and the National Science Foundation (West Lafayette, IN: Purdue University Press, 1998).

A second category of research on NSF examines the agency's practices, especially in grant-making. The classic study is Stephen Cole, Leonard Rubin, and Jonathan R. Cole, *Peer Review in the National Science Foundation: Prepared for the Committee on Science and Public Policy of the National Academy of Sciences*, 2 vols. (Washington, DC: National Academy of Sciences, 1978, 1981). Updates include George T. Mazuzan, "'Good Science Gets Funded . . . ': The Historical Evolution of Grant Making at the National Science Foundation," *Science Communication* 14 (1992): 63–90, and Marc Rothenberg, "Making Judgments about Grant Proposals: A Brief History of the Merit Review Criteria at the National Science Foundation," *Technology & Innovation* 12, no. 3 (2010): 189–95. All these studies, along with our interviews, shaped our understanding of NSF as a distinct institution with an established culture and deeply held values.

Our assessment examination of FastLane as a computing system drew heavily on the history of computing, a field that has expanded dramatically in the past 20 years. Insightful surveys include Jon Agar, *The Government Machine: A Revolutionary History of the Computer* (Cambridge, MA: MIT Press 2003); Martin Campbell-Kelly, William Aspray, Nathan Ensmenger, and Jeffery R. Yost, *Computer: A History of the Information Machine*, 3rd ed. (Boulder, CO: Westview, 2014); and Martin Campbell-Kelly and Daniel D. Garcia-Swartz, *From Mainframes to Smartphones: A History of the International Computer Industry* (Cambridge, MA: Harvard University Press, 2015). The software industry is ably covered in Martin Campbell-Kelly's *From Airline Reservations to Sonic the Hedgehog* (Cambridge, MA: MIT Press, 2003).

Studies of computing inside the federal government, beyond the military and Cold War, are as yet uncommon. For overviews, see *Funding a Revolution: Government Support for Computing Research* (Washington, DC: National Academies Press, 1999); Jon Laprise, "The Purloined Mainframe: Hiding the History of Computing in Plain Sight," *IEEE Annals of the History of Computing* 31, no. 3 (2009): 83–84; and James W. Cortada, *The Digital Hand: How Computers Changed the Work of American Public Sector Industries* (New York: Oxford University Press, 2007). Case studies include William Aspray and B. O. Williams, "Arming American Scientists: NSF and the Provision of Scientific Computing Facilities for Universities, 1950–1973," *IEEE Annals of the History of Computing* 16, no. 4 (1994): 60–74, and Janet Abbate, "Privatizing the Internet: Competing Visions and Chaotic Events, 1987–1995," *IEEE Annals of the History of Computing* 32, no. 1 (1994): 10–22. Andrew Meade McGee's "Mainframing America: Computers, Systems, and the Transformation of US Policy and Society, 1940–85" (PhD diss., University of Virginia, forthcoming) will fill a yawning gap.

The scholarly history of e-commerce and e-government is still emerging. Janet Abbate's *Inventing the Internet* (Cambridge, MA: MIT Press, 1999) and Arthur L. Norberg and Judy E. O'Neill's *Transforming Computer Technology: Information Processing for the Pentagon, 1962–1986* (Baltimore: Johns Hopkins University Press, 1996) cover the ARPANET and early Internet. A pioneering volume exploring e-commerce is William Aspray and Paul E. Ceruzzi, eds., *The Internet and American Business* (Cambridge, MA: MIT Press, 2008). For an overview of e-government, see Leon J. Osterweil, Lynette I. Millett, and Joan D. Winston, eds., *Social Security Administration Electronic Service Provision: A Strategic Assessment* (Washington, DC: National Academies Press, 2007), appendix D: "Overview of Selected Legislation" and appendix E: "A Short History of E-Government."

We read widely in the software engineering, human-centered computing, and human-computer interaction literatures. Among the insightful computer-science authors we found are Barry W. Boehm, Jonathan Grudin, and Matti Tedre, each with numerous articles.

Primary Sources

Traditional archival sources on FastLane are not plentiful. NSF typically archives only National Science Board reports and complete "jackets" on successful awards. We interviewed several FastLane principals who located dozens of primary-source documents, now in the possession of the NSF historian. Government reports by the U.S. Government Accountability Office and other agencies were valuable in understanding the wider governmental context.

Oral History Interviews

Given the thin archival resources on the history, development, and use of FastLane, we relied extensively on oral history interviews. In the formative stage of the project, we conducted nonrecorded interviews with eight individuals at NSF (Dan Atkins, Edward Hackett, Fred Kronz, Daniel Newlon, Robert O'Conner, John Perhonis, Tom Weber, and Fred Wendling), combined with two phone interviews with former NSF staff (Connie McLindon and Bruce Seely). These 10 individuals introduced us to the early planning and discussions at NSF about electronic grant submissions, as well as to the transformative contexts with FastLane's introduction and early evolution. These initial interviews gave us needed perspective and contacts to design a major oral history effort that ultimately included more than 400 in-person interviews as well as an additional 400 online interviews.

The in-person oral histories consisted of extensive interviews (generally 90 minutes to 5 hours) with key members of the FastLane design, development, and management teams, and shorter interviews (20 minutes to an hour) with three categories of FastLane users—*principal investigators* and *sponsored research personnel* at universities (29 different schools), and *NSF program officers*. NSF program officers as legacy users were involved with the external-facing FastLane platform, through interacting with PIs, and were the principal users of FastLane's internal-facing grant-making and grants-management functions (eJacket). They also provided fundamental information on the evolution to NSF's Interactive Panel System for proposal review. With the designers, developers, and managers of FastLane, we interviewed multiple individuals on key developments and issues in an effort to triangulate and gain accurate information. For the oral histories with FastLane users, large scale and intentional diversity helped to ensure quality data.

To further broaden the dataset on FastLane users, and to test qualitative aspects of electronic interviews, the project team developed an online interview platform that resulted in roughly 400 additional interviews with PIs and sponsored research personnel from universities. Our research team was interested in understanding (for this project and potential future ones on IT users) the effectiveness of such online interviewing technology in securing quality data. As a quasi-experiment, individuals from the same schools visited for in-person interviews were invited (through email) to complete the online interviews. A small number of additional online respondents happened on the interview platform and independently completed the online interview. The same basic questions were asked in the online interviews of university FastLane users as in the in-person interviews, but unlike the in-person interviews, the online platform did not offer dynamic communication and follow-up, "off-script" questions and conversations. We also tended to get longer, more reflective, responses with the in-person interviews. This highlights the qualitative value for in-person oral history interviews—versus surveys or online interview platforms—even for shorter interviews with users of IT systems.

The interviews we conducted with members of the FastLane design, development, and management teams were similar in form to the research-grade oral histories that Charles Babbage Institute historians have been conducting for more than three decades, and for which both authors have many years of interviewing experience. They involved targeted preparation before each interview. Oral histories with the Division of Information System's Connie McLindon and Fred Wendling provide insights into FastLane's origins and offer underlying context to the archival documents obtained, such as the Weber report. An interview with FastLane software contractor Rich Schneider conveys valuable data from one of the key individuals who programmed a substantial portion of the code on the original system. Interviews with FastLane DIS project managers Carolyn Miller and Craig Robinson offer insight on the management of the project through its rollout and early refinement. Longtime DIS professional Beverly Sherman was one of the two principal individuals who traveled to universities and professional meetings (the Society of Research Administrators and the National Council of University Research Administrators) to hold FastLane workshops and introduce the system to administrators and those in the research community. Her interview illustrates how NSF staff continually listened to and took cues on design and refinement of the system from stakeholders. An interview with DIS director Andrea Norris gives important insights on the timing and funding challenges with rolling out eJacket in different phases. Longtime leader of the NSF Policy Office, Jean Feldman, conveys the symbiotic relationship of policy and IT as FastLane and eJacket were designed, introduced, and modified. Our interview with Paul Morris characterizes the backstory of user-driven innovation with his widely used automated grant compliance checker. A senior NSF human resources executive, Joseph Burt, helped us understand the social and managerial aspects of FastLane and eJacket and successes and challenges with changing work roles internally, while Erika Rissi's oral history is invaluable to understanding the early work on FastLane's modular successor, Research.gov. Some interview candidates arose organically from their identification in prior interviews while others were the product of our document research. Nearly all the designers, developers, and managers at NSF whom we approached consented to interviews.

Likewise, dozens of program officers at NSF were generous with their time in agreeing to participate. In all, we conducted 48 in-person interviews with program officers. These tended to be roughly an hour in length and covered their experience with the introduction of FastLane, the different phases of the IT-enabled interactive panel review, and the launch of eJacket, first for proposal declinations, and then, years later, for awards. While we prepared core questions, there was great value to going off script with this community. Most program officer interviewees were permanent employees with 15 or more years of service at NSF. In selecting interview candidates, we wanted program officers with experience both in the paper-based days and with FastLane eJacket. Most temporary rotators lack the pre-FastLane experience at NSF, though a few had returned after an earlier tour of duty and were interviewed.

For the sponsored research administrators and PI interviews, we visited 29 universities. Some trips we took together, although most were done individually by one of the two co-authors. Diversity was a fundamental consideration in selecting the schools we visited and the people we invited to participate. On the advice of NSF historian Marc Rothenberg and a few other NSF veterans, we intentionally oversampled EPSCoR states (12 schools) and historically black colleges and universities (5 schools). This was to effectively gauge whether the introduction of new submission requirements and technology with FastLane provided greater hardship for schools with fewer institutional resources. We gathered baseline data by interviewing at schools of various sizes and types—public and private, secular and religious-based, urban and rural / small city—regardless of EPSCoR status. Overall, we conducted in-person interviews with hundreds of PIs and dozens of sponsored research administrators—ranging from entry-level grants specialists to directors of sponsored research and vice presidents for research.

Diversity was also at the forefront of our selection with regard to geography, gender, researchers' discipline, and seniority. We visited multiple schools in the Northwest, Southwest, Midwest, Northeast, and Southeast, as well as Alaska, Hawaii, the U.S. Virgin Islands, and Puerto Rico. Schools visited are listed in appendix A. In selecting whom to invite, we sought interviews with all sponsored research office personnel who had both pre-FastLane (paper-based days) and FastLane experience. This same experience (paper and FastLane) was sought in selecting PIs. While some junior faculty were interviewed, senior scholars (associate and full professors) were more likely to have paper-based and FastLane as part of their experience with NSF grant submissions. The median PI interviewee was 24 years out from receiving a doctorate. Summary statistics on interviewees are provided in appendix B.

We identified potential PI interviewees using the NSF's website—the awarded grants database. While this created a selection bias for people who had at least one awarded NSF project, many interviewees recounted both successes and shortcomings with receiving NSF funding, and we do not believe this meaningfully skewed results. We sought to assure gender diversity of interviewees; more than 40 percent of our in-person interviewees were with women. Different areas of the natural sciences, life sciences, engineering, and social sciences have different practices with proposing research and disseminating results—for instance, the use of color graphics. This was one of the key reasons we also sought diversity with regard to discipline and had a substantial number of interviewees in biology, chemistry, computer science, the earth sciences, engineering, mathematics, physics, and the social sciences.

The in-person interviews for users at universities started with about 20 questions, but we always looked for opportunities to pursue follow-up lines of inquiry. For PIs, we sought to learn about their experiences with grant submission pre-FastLane (paper-based days), initial experiences with the user interface and functionality of the FastLane system, available networking and software resources (Netscape and Adobe Acrobat in the early days), institutional support, impressions on changes with FastLane over time, and lessons from FastLane about the design and deployment of cyberinfrastructures.

Interviews with sponsored research personnel tended to focus on their own (often daily) interaction with the system in providing institutional support and submission authority. We were particularly concerned with how these sponsored research offices implemented changes in policies and practices with the introduction and evolution of FastLane. There was more to explore on the pre-award side, so we interviewed more staff focused in this area. However, post-award specialists were interviewed as well and provided information on how the system was used to draw funds on awarded projects. Many interviewees compared and contrasted the features and usability of FastLane and Grants.gov, generally emphasizing their great preference for FastLane.

Index

Page numbers in *italics* refer to illustrations.